Calculus II

Mehdi Rahmani-Andebili

Calculus II

Practice Problems, Methods, and Solutions

 Springer

Mehdi Rahmani-Andebili
Electrical Engineering Department
Arkansas Tech University
Russellville, AR, USA

ISBN 978-3-031-45352-6   ISBN 978-3-031-45353-3   (eBook)
https://doi.org/10.1007/978-3-031-45353-3

This Springer imprint is published by the registered company Springer Nature Switzerland AG
The registered company address is: Gewerbestrasse 11, 6330 Cham, Switzerland

Paper in this product is recyclable.

Preface

Calculus is one of the most important courses of many majors, including engineering and science, and even some non-engineering majors like economics and business, which is taught in three successive courses at universities and colleges worldwide. Moreover, in many universities and colleges, a precalculus course is mandatory for under-prepared students as the prerequisite course of Calculus 1.

Unfortunately, some students do not have a solid background and knowledge in math and calculus when they start their education in universities or colleges. This issue prevents them from learning calculus-based courses such as physics and engineering. Sometimes, the problem escalates, so they give up and leave the university. Based on my real professorship experience, students do not have a serious issue comprehending physics and engineering courses. In fact, it is the lack of enough knowledge of calculus that hinder them from understanding those courses.

Therefore, a series of calculus textbooks covering Precalculus, Calculus 1, Calculus 2, and Calculus 3 have been prepared to help students succeed in their major. The subjects of the calculus series books are as follows.

Precalculus: Practice Problems, Methods, and Solution

- *Real Number Systems, Exponents and Radicals, and Absolute Values and Inequalities*
- *Systems of Equations*
- *Quadratic Equations*
- *Functions, Algebra of Functions, and Inverse Functions*
- *Factorization of Polynomials*
- *Trigonometric and Inverse Trigonometric Functions*
- *Arithmetic and Geometric Sequences*

Calculus 1: Practice Problems, Methods, and Solution

- *Characteristics of Functions*
- *Trigonometric Equations and Identities*
- *Limits and Continuities*
- *Derivatives and Their Applications*
- *Definite and Indefinite Integrals*

Calculus 2: Practice Problems, Methods, and Solution

- *Applications of Integration*
- *Sequences and Series and Their Applications*
- *Polar Coordinate System*
- *Complex Numbers*

Calculus 3: Practice Problems, Methods, and Solution

- *Linear Algebra and Analytical Geometry*
- *Lines, Surfaces, and Vector Functions in Three-Dimensional Coordinate System*
- *Multivariable Functions*
- *Double Integrals and their Applications*
- *Triple Integrals and their Applications*
- *Line Integrals and their Applications*

The textbooks include basic and advanced calculus problems with very detailed problem solutions. They can be used as practicing study guides by students and as supplementary teaching sources by instructors. Since the problems have very detailed solutions, the textbooks are helpful for under-prepared students. In addition, they are beneficial for knowledgeable students because they include advanced problems.

In preparing the problems and solutions, care has been taken to use methods typically found in the primary instructor-recommended textbooks. By considering this key point, the textbooks are in the direction of instructors' lectures, and the instructors will not see any untaught and unusual problem solutions in their students' answer sheets.

To help students study in the most efficient way, the problems have been categorized into nine different levels. In this regard, for each problem, a difficulty level (easy, normal, or hard) and a calculation amount (small, normal, or large) have been assigned. Moreover, problems have been ordered in each chapter from the easiest problem with the smallest calculations to the most difficult problems with the largest ones. Therefore, students are suggested to start studying the textbooks from the easiest problems and continue practicing until they reach the normal and then the hardest ones. This classification can also help instructors choose their desirable problems to conduct a quiz or a test. Moreover, the classification of computation amount can help students manage their time during future exams, and instructors assign appropriate problems based on the exam duration.

Russellville, AR, USA   Mehdi Rahmani-Andebili

The Other Works Published by the Author

The author has already published the books and textbooks below with Springer Nature.

Precalculus (2nd Ed.) – Practice Problems, Methods, and Solutions, *Springer Nature*, 2023.

Calculus III – Practice Problems, Methods, and Solutions, *Springer Nature*, 2023.

Calculus II – Practice Problems, Methods, and Solutions, *Springer Nature*, 2023.

Calculus I (2nd Ed.) – Practice Problems, Methods, and Solutions, *Springer Nature*, 2023.

Planning and Operation of Electric Vehicles in Smart Grid, *Springer Nature*, 2023.

Applications of Artificial Intelligence in Planning and Operation of Smart Grid, *Springer Nature*, 2022.

AC Electric Machines- Practice Problems, Methods, and Solutions, *Springer Nature*, 2022.

DC Electric Machines, Electromechanical Energy Conversion Principles, and Magnetic Circuit Analysis- Practice Problems, Methods, and Solutions, *Springer Nature*, 2022.

Differential Equations- Practice Problems, Methods, and Solutions, *Springer Nature*, 2022.

Feedback Control Systems Analysis and Design- Practice Problems, Methods, and Solutions, *Springer Nature*, 2022.

Power System Analysis – Practice Problems, Methods, and Solutions, *Springer Nature*, 2022.

Advanced Electrical Circuit Analysis – Practice Problems, Methods, and Solutions, *Springer Nature*, 2022.

Design, Control, and Operation of Microgrids in Smart Grids, *Springer Nature*, 2021.

Applications of Fuzzy Logic in Planning and Operation of Smart Grids, *Springer Nature*, 2021.

Operation of Smart Homes, *Springer Nature*, 2021.

AC Electrical Circuit Analysis – Practice Problems, Methods, and Solutions, *Springer Nature*, 2021.

Calculus – Practice Problems, Methods, and Solutions, *Springer Nature*, 2021.

Precalculus – Practice Problems, Methods, and Solutions, *Springer Nature*, 2021.

DC Electrical Circuit Analysis – Practice Problems, Methods, and Solutions, *Springer Nature*, 2020.

Planning and Operation of Plug-in Electric Vehicles: Technical, Geographical, and Social Aspects, *Springer Nature*, 2019.

Contents

Abstract

In this chapter, the basic and advanced problems related to the applications of integration are presented. The subjects include mean value of a function, surface area bounded by curves, volume resulted from rotation of an enclosed region, arc length of a curve, surface area of a solid of revolution, and center of gravity. In this chapter, the problems are categorized in different levels based on their difficulty levels (easy, normal, and hard) and calculation amounts (small, normal, and large). Additionally, the problems are ordered from the easiest problem with the smallest computations to the most difficult problems with the largest calculations.

1.1 Mean Value of a Function

1.1. For the range of $2 \leq x \leq 5$, calculate the mean value of the following function [1–3].

$$y = ax + b$$

Difficulty level   ● Easy   ○ Normal   ○ Hard
Calculation amount   ● Small   ○ Normal   ○ Large

1) $\frac{5}{2}a + 3b$

2) $\frac{7}{2}a + 3b$

3) $\frac{5}{2}a + b$

4) $\frac{7}{2}a + b$

1.2. Consider the functions of $f(x) = 2x$ and $g(x) = 3x^2 - 2x$. Calculate the value of λ if the mean value of the functions in the range of $[1, \lambda]$ is the same.

Difficulty level   ● Easy   ○ Normal   ○ Hard
Calculation amount   ● Small   ○ Normal   ○ Large

1) $\frac{1 + \sqrt{5}}{2}$

2) $\frac{1 + \sqrt{3}}{2}$

3) $\frac{\sqrt{5}}{2}$

4) $\frac{3\sqrt{3}}{2}$

M. Rahmani-Andebili, *Calculus II*, https://doi.org/10.1007/978-3-031-45353-3_1

1.2   Surface Area Bounded by Curves

1.3. Calculate the surface area enclosed between the curves of $y = 2x^2 - 2x$ and $y = x^2$.

Difficulty level   ● Easy   ○ Normal   ○ Hard
Calculation amount   ● Small   ○ Normal   ○ Large

1) $\dfrac{1}{3}$

2) $\dfrac{2}{3}$

3) $\dfrac{4}{3}$

4) $\dfrac{7}{3}$

1.4. Calculate the surface area bounded by the functions of $y(x) = \sin x$ and $y(x) = \cos x$ for $x = \left[\dfrac{\pi}{4}, \dfrac{5\pi}{4}\right]$.

Difficulty level   ● Easy   ○ Normal   ○ Hard
Calculation amount   ● Small   ○ Normal   ○ Large

1) $\dfrac{1}{2}$

2) $\sqrt{2}$

3) $2\sqrt{2}$

4) $3\sqrt{2}$

1.5. Calculate the surface area enclosed between the curves of $y = x^2$ and $y = \sqrt{x}$.

Difficulty level   ○ Easy   ● Normal   ○ Hard
Calculation amount   ● Small   ○ Normal   ○ Large

1) $\dfrac{2}{3}$

2) 1

3) $\dfrac{1}{3}$

4) $\dfrac{1}{6}$

1.6. Calculate the surface area enclosed between the curve of $y = x^3 + 2x^2 + x$ and x-axis.

Difficulty level   ○ Easy   ● Normal   ○ Hard
Calculation amount   ● Small   ○ Normal   ○ Large

1) $\dfrac{1}{12}$

2) $\dfrac{1}{10}$

3) $\dfrac{1}{9}$

4) $\dfrac{1}{7}$

1.7. Calculate the surface area enclosed between the curve of $y = x^2 + 1$ and the line of $y = 2$.

Difficulty level   ○ Easy   ● Normal   ○ Hard
Calculation amount   ● Small   ○ Normal   ○ Large

1) $\dfrac{1}{3}$

2) $\dfrac{2}{3}$

3) 1

4) $\dfrac{4}{3}$

1.8. Calculate the surface area restricted by the curve of $y^2(x) = x^2 - x^4$.

Difficulty level   ○ Easy   ● Normal   ○ Hard
Calculation amount   ● Small   ○ Normal   ○ Large

1) 1

2) 2

3) $\dfrac{2}{3}$

4) $\dfrac{4}{3}$

1.9. Calculate the surface area of the shaded region shown in Fig. 1.1.

Difficulty level   ○ Easy   ● Normal   ○ Hard
Calculation amount   ● Small   ○ Normal   ○ Large

1) $8 - 2\pi$

2) $8 - \dfrac{\pi}{2}$

3) $4 - \dfrac{\pi}{4}$

4) $2\pi - 4$

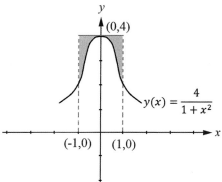

Figure 1.1 The graph of problem 1.9

1.10. Calculate the surface area restricted by the following function, above the line of $y(x) = 0$, and the right-hand side of $x = 1$.

$$y(x) = \frac{6}{(2x+1)(x+2)}$$

Difficulty level   ○ Easy   ● Normal   ○ Hard
Calculation amount   ● Small   ○ Normal   ○ Large

1) $\dfrac{1}{2}$

2) ln 2

3) 2 ln 2

4) ∞

1.11. Calculate the surface area enclosed between the curve with the function below and x-axis in the range of $\left[-\sqrt{2}, \sqrt{2}\right]$.

$$y = \frac{1}{2 + x^2}$$

Difficulty level   ○ Easy   ● Normal   ○ Hard
Calculation amount   ○ Small   ● Normal   ○ Large

1) $\dfrac{\pi\sqrt{2}}{4}$

2) $\dfrac{\pi}{4}$

3) $\dfrac{\pi\sqrt{2}}{2}$

4) $\dfrac{\pi}{2}$

1.12. Calculate the surface area restricted by the function below and x-axis in the domain of $[1, e^2]$.

$$y(x) = \frac{\ln x}{\sqrt{x}}$$

Difficulty level   ○ Easy   ● Normal   ○ Hard
Calculation amount   ○ Small   ● Normal   ○ Large

1) 5
2) 4
3) 3
4) 2

1.13. Calculate the surface area of the shaded region shown in Fig. 1.2.

Difficulty level   ○ Easy   ● Normal   ○ Hard
Calculation amount   ○ Small   ● Normal   ○ Large

1) $\dfrac{1}{3}$

2) $\dfrac{2}{3}$

3) $\dfrac{1}{\ln 2} - \dfrac{2}{3}$

4) $\dfrac{2}{3} - \ln 2$

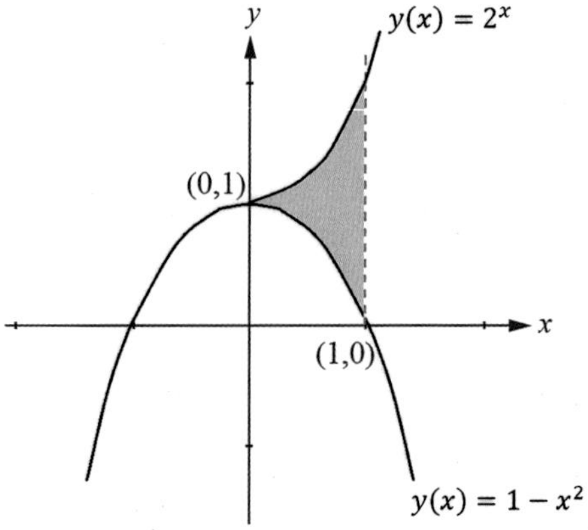

Figure 1.2 The graph of problem 1.13

1.14. Calculate the surface area of the shaded region shown in Fig. 1.3.

Difficulty level   ○ Easy   ● Normal   ○ Hard
Calculation amount   ○ Small   ● Normal   ○ Large

1) π

2) $\dfrac{\pi}{2}$

3) $\dfrac{\pi}{2} - \dfrac{1}{2}$

4) $\dfrac{\pi}{2} + \dfrac{1}{2}$

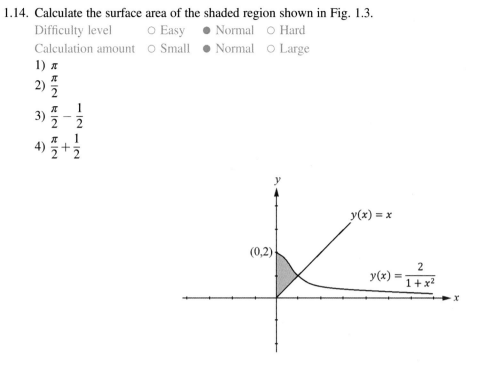

Figure 1.3 The graph of problem 1.14

1.15. Determine the value of parameter of c, where $0 < c < \dfrac{\pi}{2}$, if the surface area restricted by the function of $y(x) = \cos x$, the function of $y(x) = \cos(x - c)$, and the line of $x = 0$ is equal to the surface area bounded by the function of $y(x) = \cos(x - c)$, the line of $x = \pi$, and the line of $y(x) = 0$.

Difficulty level   ○ Easy   ● Normal   ○ Hard
Calculation amount   ○ Small   ○ Normal   ● Large

1) $\dfrac{\pi}{5}$

2) $\dfrac{\pi}{3}$

3) $\dfrac{\pi}{4}$

4) $\dfrac{\pi}{6}$

1.16. Determine the surface area bounded by the function of $y(x) = x \ln \sqrt{x}$, x-axis, and the lines with the equations of $x = 1$ and $x = 2$.

Difficulty level   ○ Easy   ○ Normal   ● Hard
Calculation amount   ● Small   ○ Normal   ○ Large

1) $2 \ln 2 - \dfrac{1}{2}$

2) $2 \ln 2 - 1$

3) $\ln 2 + \dfrac{3}{4}$

4) $\ln 2 - \dfrac{3}{8}$

1.17. Estimate the surface area of the shaded region shown in Fig. 1.4.

Difficulty level   ○ Easy   ○ Normal   ● Hard
Calculation amount   ○ Small   ● Normal   ○ Large

1) ∞

2) $\ln 2$

3) cos2
4) 2 ln 2 − cos 2

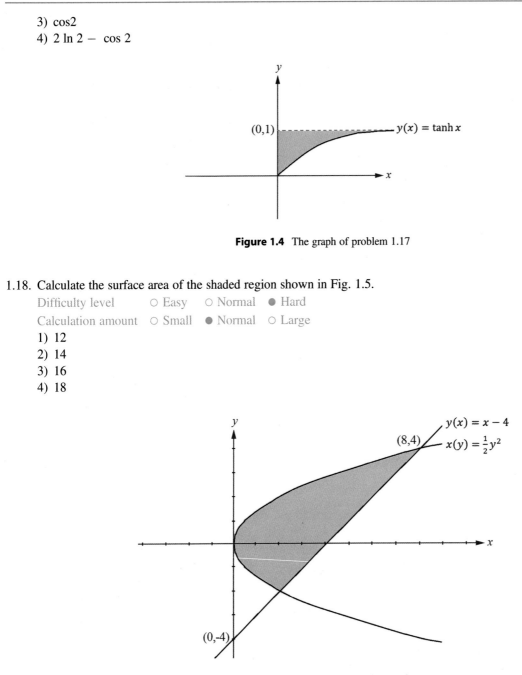

Figure 1.4 The graph of problem 1.17

1.18. Calculate the surface area of the shaded region shown in Fig. 1.5.

Difficulty level   ○ Easy   ○ Normal   ● Hard
Calculation amount   ○ Small   ● Normal   ○ Large

1) 12
2) 14
3) 16
4) 18

Figure 1.5 The graph of problem 1.18

1.3   Volume Resulted from Rotation of an Enclosed Region

1.19. Calculate the volume resulted from the rotation of the function of $y(x) = \sqrt{x^2 - x^3}$ around x-axis for $x = [-1, 1]$.

Difficulty level   ● Easy   ○ Normal   ○ Hard
Calculation amount   ● Small   ○ Normal   ○ Large

1) $\frac{\pi}{3}$

2) $\frac{2\pi}{3}$

3) π

4) $\dfrac{3\pi}{2}$

1.20. Calculate the volume resulted from the rotation of the surface area around x-axis enclosed between one period of the curve of $y = \sin(x)$ and x-axis.

Difficulty level   ○ Easy   ● Normal   ○ Hard

Calculation amount   ● Small   ○ Normal   ○ Large

1) π^2

2) $2\pi^2$

3) $\dfrac{\pi^2}{2}$

4) $\dfrac{\pi^2}{4}$

1.21. Calculate the volume resulted from the rotation of the function of $y(x) = \cos x$ around x-axis for $x = \left[0, \dfrac{\pi}{2}\right]$.

Difficulty level   ○ Easy   ● Normal   ○ Hard

Calculation amount   ● Small   ○ Normal   ○ Large

1) $\dfrac{\pi^2}{4}$

2) $\dfrac{\pi^2}{8}$

3) $\dfrac{\pi}{8}$

4) $\dfrac{\pi}{4}$

1.22. The volume resulted from the rotation of the surface area restricted by the function of $y(x) = e^{-x}$, the coordinate axes, and $x = b$, where $b > 0$, around x-axis is V. Calculate the value of $\lim\limits_{b \to \infty} V$.

Difficulty level   ○ Easy   ● Normal   ○ Hard

Calculation amount   ● Small   ○ Normal   ○ Large

1) $\dfrac{\pi}{4}$

2) $\dfrac{\pi}{2}$

3) π

4) ∞

1.23. The surface area restricted by the function of $y(x) = \dfrac{1}{\sqrt{x}}$, the coordinate axes, and $x = 1$ is called S. Moreover, the volume resulted from the rotation of the surface area around x-axis is called V. Which of the following options is correct?

Difficulty level   ○ Easy   ● Normal   ○ Hard

Calculation amount   ○ Small   ● Normal   ○ Large

1) $S = 2, V = 8\pi$

2) $S = 4, V = 8\pi$

3) $S = 2, V = \infty$

4) $S = \infty, V = \infty$

1.24. Calculate the volume resulted from the rotation of the surface area around y-axis enclosed between the curve of $y = 1 - \dfrac{1}{4}x^2$ and x-axis.

Difficulty level   ○ Easy   ○ Normal   ● Hard

Calculation amount   ○ Small   ● Normal   ○ Large

1) π

2) 2π

3) 3π

4) 4π

1.25. Calculate the volume resulted from the rotation of the surface area around x-axis enclosed between the curve with the function below, x-axis, $x = \frac{\pi}{4}$, and $x = \frac{\pi}{2}$.

$$y = \frac{1}{\sin^2(x)}$$

Difficulty level   ○ Easy   ○ Normal   ● Hard

Calculation amount   ○ Small   ● Normal   ○ Large

1) π

2) $\frac{2\pi}{3}$

3) 2π

4) $\frac{4\pi}{3}$

1.26. The surface area confined by the function below, x-axis, and two lines with the equations of $x = \frac{\pi}{6}$ and $x = \frac{\pi}{2}$ are rotated around x-axis. Calculate the resultant volume.

$$y(x) = \frac{\sqrt{\cos^3 x}}{\sin^2 x}$$

Difficulty level   ○ Easy   ○ Normal   ● Hard

Calculation amount   ○ Small   ● Normal   ○ Large

1) $\frac{7\pi}{3}$

2) $\frac{5\pi}{3}$

3) $\frac{4\pi}{3}$

4) $\frac{2\pi}{3}$

1.27. Calculate the volume resulted from the rotation of the surface area bounded by the function of $y(x) = xe^x$ and the lines with the equations of $x = 1$ and $y(x) = 0$ around x-axis.

Difficulty level   ○ Easy   ○ Normal   ● Hard

Calculation amount   ○ Small   ● Normal   ○ Large

1) $\frac{\pi}{2}(e^2 - 1)$

2) $\frac{\pi}{2}(e^2 - 2)$

3) $\frac{\pi}{4}(e^2 - 1)$

4) $\frac{\pi}{4}(e^2 + 1)$

1.28. Calculate the volume resulted from the rotation of the surface area restricted by the relation of $y^{\frac{1}{2}}(x) = a^{\frac{1}{2}} - x^{\frac{1}{2}}$ and x- and y- axes around x-axis.

Difficulty level   ○ Easy   ○ Normal   ● Hard

Calculation amount   ○ Small   ● Normal   ○ Large

1) $5\pi a^3$

2) $\frac{1}{2}\pi a^3$

3) $\frac{1}{12}\pi a^3$

4) $\frac{1}{15}\pi a^3$

1.29. Calculate the volume resulted from the rotation of the function of $y(x) = e^{-x}\sqrt{\sin x}$ around x-axis for $0 \le x \le \pi$.

Difficulty level ○ Easy ○ Normal ● Hard
Calculation amount ○ Small ○ Normal ● Large

1) $\frac{\pi}{5}\left(\frac{1}{1 - e^{-2\pi}}\right)$

2) $\frac{\pi}{5}\left(\frac{1}{1 + e^{-2\pi}}\right)$

3) $\frac{\pi}{5}\left(1 + e^{-2\pi}\right)$

4) $\frac{\pi}{5}\left(1 - e^{-2\pi}\right)$

1.30. Calculate the volume created by the rotation of the shaded region (see Fig. 1.6) around x-axis.

Difficulty level ○ Easy ○ Normal ● Hard
Calculation amount ○ Small ○ Normal ● Large

1) $\frac{4\pi}{15}$

2) $\frac{\pi}{2}$

3) $\frac{14\pi}{3}$

4) $\frac{3\pi}{5}$

Figure 1.6 The graph of problem 1.30

1.4   Arc Length of a Curve

1.31. Calculate the arc length of the function of $f(x) = \int_1^x \sqrt{t^4 - 1}\, dt$ for the interval of $x = [1, 3]$.

Difficulty level   ○ Easy   ● Normal   ○ Hard
Calculation amount   ● Small   ○ Normal   ○ Large

1) $\dfrac{19}{3}$

2) $\dfrac{25}{3}$

3) $\dfrac{26}{3}$

4) $\dfrac{28}{3}$

1.32. Calculate the arc length of the parametric relation below for the interval of $t = [0, 4]$.

$$\begin{cases} x(t) = e^t \cos t \\ y(t) = e^t \sin t \end{cases}$$

Difficulty level   ○ Easy   ● Normal   ○ Hard
Calculation amount   ● Small   ○ Normal   ○ Large

1) $\sqrt{2}(e^4 - 1)$
2) $\sqrt{2}(e^4 + 1)$
3) $2(e^4 - 1)$
4) $2(e^4 + 1)$

1.33. Calculate the arc length of the following parametric relation for the interval of $0 \le t \le 1$.

$$\begin{cases} x(t) = e^t(\cos t + \sin t) \\ y(t) = e^t(\cos t - \sin t) \end{cases}$$

Difficulty level   ○ Easy   ● Normal   ○ Hard
Calculation amount   ○ Small   ● Normal   ○ Large

1) $e - 1$
2) $\dfrac{e - 1}{2}$
3) $4(e - 1)$
4) $2(e - 1)$

1.34. Calculate the arc length of the function of $y(x) = \dfrac{1}{2}x^2 - \dfrac{1}{4}\ln x$ for the interval of $x = [1, 2]$.

Difficulty level   ○ Easy   ● Normal   ○ Hard
Calculation amount   ○ Small   ● Normal   ○ Large

1) $\dfrac{3}{2} + \dfrac{1}{2}\ln 2$
2) $\dfrac{3}{2} + \dfrac{1}{4}\ln 2$
3) $\dfrac{3}{4} + \dfrac{1}{2}\ln 2$
4) $\dfrac{3}{4} + \dfrac{1}{4}\ln 2$

1.35. Calculate the arc length of the function of $y(x) = \dfrac{x^4}{8} + \dfrac{1}{4x^2}$ for the interval of $x = [1, 2]$.

Difficulty level   ○ Easy   ● Normal   ○ Hard
Calculation amount   ○ Small   ● Normal   ○ Large

1) 15

2) 21

3) $\dfrac{25}{3}$

4) $\dfrac{33}{16}$

1.36. Calculate the arc length of the function of $f(x) = \ln(\sec x)$ for the interval of $x = \left[0, \dfrac{\pi}{4}\right]$.

Difficulty level   ○ Easy   ○ Normal   ● Hard

Calculation amount   ○ Small   ● Normal   ○ Large

1) $\ln\sqrt{2}$

2) $\ln\left|2 - \sqrt{2}\right|$

3) $\ln\left(\sqrt{2} - 1\right)$

4) $\ln\left(1 + \sqrt{2}\right)$

1.37. Calculate the arc length of the function of $f(x) = \displaystyle\int_0^x \sqrt{\cosh(t)}\,dt$ for the interval of $x = [0, 2]$.

Difficulty level   ○ Easy   ○ Normal   ● Hard

Calculation amount   ○ Small   ● Normal   ○ Large

1) $\sqrt{2}\left(e - \dfrac{1}{e}\right)$

2) $2\left(e - \dfrac{1}{e}\right)$

3) $\sqrt{2}\left(e + \dfrac{1}{e}\right)$

4) $2\left(e + \dfrac{1}{e}\right)$

1.38. Calculate the arc length of the function of $f(x) = \dfrac{1}{2}\left[x\sqrt{x^2 - 1} - \ln\left(x + \sqrt{x^2 - 1}\right)\right]$ for the interval of $x = [1, 2]$.

Difficulty level   ○ Easy   ○ Normal   ● Hard

Calculation amount   ○ Small   ● Normal   ○ Large

1) $\dfrac{1}{2}$

2) $\dfrac{3}{2}$

3) $1 + \sqrt{2}$

4) 2

1.39. Calculate the arc length of the function of $f(x) = \ln\left(\dfrac{e^x + 1}{e^x - 1}\right)$ for the interval of $0 \leq x \leq 2$.

Difficulty level   ○ Easy   ○ Normal   ● Hard

Calculation amount   ○ Small   ○ Normal   ● Large

1) $\ln\left(e - \dfrac{1}{e}\right)$

2) $\ln\left(e + \dfrac{1}{e}\right)$

3) $\ln\left(e^2 - \dfrac{1}{e^2}\right)$

4) $\ln\left(e^2 + \dfrac{1}{e^2}\right)$

1.5   Surface Area of a Solid of Revolution

1.40. The function of $y(x) = \cosh x$ for the interval of $0 \leq x \leq 2$ is rotated around x-axis. Calculate the surface area of the solid.

Difficulty level   ○ Easy   ● Normal   ○ Hard
Calculation amount   ○ Small   ● Normal   ○ Large

1) $\dfrac{\pi}{2}(2 + \cosh 2)$

2) $\dfrac{\pi}{2}(2 + \sinh 2)$

3) $\dfrac{\pi}{2}(4 + \cosh 4)$

4) $\dfrac{\pi}{2}(4 + \sinh 4)$

1.41. The function of $3y(x) - x^3 = 0$ is rotated around x-axis. Calculate the surface area of the solid of revolution.

Difficulty level   ○ Easy   ● Normal   ○ Hard
Calculation amount   ○ Small   ● Normal   ○ Large

1) $\dfrac{\pi}{9}\left(2\sqrt{2} - 1\right)$

2) $\dfrac{\pi}{9}\left(\sqrt{2} - 1\right)$

3) $\dfrac{\pi}{3}\left(\sqrt{2} - 1\right)$

4) $\dfrac{\pi}{3}\left(2\sqrt{2} - 1\right)$

1.42. The function of $y(x) = x^2$ for the interval of $y \leq 2$ is rotated around y-axis. Calculate the surface area of the solid of revolution.

Difficulty level   ○ Easy   ○ Normal   ● Hard
Calculation amount   ○ Small   ● Normal   ○ Large

1) $\dfrac{8\pi}{3}$

2) $\dfrac{13\pi}{3}$

3) $\dfrac{11\pi}{3}$

4) $\dfrac{10\pi}{3}$

1.43. The function of $f(x) = 6\cosh \dfrac{x}{6}$ for the interval of $0 \leq x \leq 5$ is rotated around x-axis. What is the ratio of the volume of the solid of revolution to its surface area.

Difficulty level   ○ Easy   ○ Normal   ● Hard
Calculation amount   ○ Small   ● Normal   ○ Large

1) 3

2) 2

3) $\dfrac{3}{2}$

4) $\dfrac{2\pi}{3}$

1.44. The function of $f(x) = \int_0^x \sinh(t^2)\,dt$ for the interval of $0 \le x \le \sqrt{\ln 1396}$ is rotated around y-axis. Calculate the surface area of the solid of revolution.

Difficulty level   ○ Easy   ○ Normal   ● Hard
Calculation amount   ○ Small   ● Normal   ○ Large

1) $2\pi\left(1396 - \dfrac{1}{1396}\right)$

2) $\dfrac{\pi}{4}\left(1396 - \dfrac{1}{1396}\right)$

3) $\pi\left(1396 - \dfrac{1}{1396}\right)$

4) $\dfrac{\pi}{2}\left(1396 - \dfrac{1}{1396}\right)$

1.6   Center of Gravity

1.45. Determine the center of gravity of a surface from x-axis which is restricted by the function of $f(x) = 1 - x^2$ and x-axis.

Difficulty level   ○ Easy   ● Normal   ○ Hard
Calculation amount   ○ Small   ● Normal   ○ Large

1) $\dfrac{4}{15}$

2) $\dfrac{4}{5}$

3) $\dfrac{3}{5}$

4) $\dfrac{2}{5}$

1.46. Determine the center of gravity of a surface from y-axis which is restricted by the function of $f(x) = 1 - x^2$ and x-axis.

Difficulty level   ○ Easy   ● Normal   ○ Hard
Calculation amount   ○ Small   ● Normal   ○ Large

1) 0

2) $\dfrac{4}{5}$

3) $\dfrac{3}{5}$

4) $\dfrac{2}{5}$

1.47. Determine the center of gravity of the parametric curve below from x-axis.

$$\begin{cases} x(t) = t + \sin t \\ y(t) = 1 - \cos t \end{cases}, \quad 0 \le t \le \pi$$

Difficulty level   ○ Easy   ○ Normal   ● Hard
Calculation amount   ○ Small   ○ Normal   ● Large

1) 1

2) $\dfrac{1}{3}$

3) $\dfrac{2}{3}$

4) $\dfrac{3}{4}$

References

1. Rahmani-Andebili, M. (2023). Calculus I (2nd Ed.) – Practice Problems, Methods, and Solutions, Springer Nature, 2021.
2. Rahmani-Andebili, M. (2021). Calculus – Practice Problems, Methods, and Solutions, Springer Nature, 2021.
3. Rahmani-Andebili, M. (2021). Precalculus – Practice Problems, Methods, and Solutions, Springer Nature, 2021.

2

Abstract

In this chapter, the problems of the first chapter are fully solved, in detail, step-by-step, and with different methods.

2.1   Mean Value of a Function

2.1. As we know, the average value of a function can be determined as follows [1–3]:

$$f_{ave} = \frac{1}{b-a} \int_a^b f(x)dx$$

In addition, from list of integral of functions, we know that:

$$\int x^n dx = \frac{1}{n+1} x^{n+1} + c$$

The problem can be solved as follows.

$$f_{ave} = \frac{1}{5-2} \int_2^5 (ax+b)dx = \frac{1}{3} \left(\frac{a}{2} x^2 + bx \right) \Big|_2^5 = \frac{1}{3} \left(\frac{25a}{2} + 5b - \frac{4a}{2} - 2b \right)$$

$$\Rightarrow f_{ave} = \frac{7}{2} a + b$$

Choice (4) is the answer.

2.2. As we know, the average value of a function can be determined as follows:

$$\frac{1}{b-a} \int_a^b f(x)dx$$

Based on the problem, we know that:

$$f_{ave} = g_{ave}$$

Now, the problem can be solved as follows.

M. Rahmani-Andebili, *Calculus II*, https://doi.org/10.1007/978-3-031-45353-3_2

$$\frac{1}{\lambda - 1} \int_1^\lambda 2x \, dx = \frac{1}{\lambda - 1} \int_1^\lambda (3x^2 - 2x) \, dx$$

$$\Rightarrow x^2 \Big|_1^\lambda = (x^3 - x^2) \Big|_1^\lambda$$

$$\Rightarrow \lambda^2 - 1 = (\lambda^3 - \lambda^2) - 0 \Rightarrow \lambda^3 - 2\lambda^2 + 1 = 0$$

$$\Rightarrow (\lambda - 1)(\lambda^2 - \lambda - 1) = 0$$

$$\Rightarrow \lambda = \frac{1 - \sqrt{5}}{2}, \frac{1 + \sqrt{5}}{2}, 1$$

However, just $\frac{1+\sqrt{5}}{2}$ is acceptable because the others are not within the range.

$$\Rightarrow \lambda = \frac{1 + \sqrt{5}}{2}$$

Choice (1) is the answer.

In this problem, the rule below was used.

$$\int x^n \, dx = \frac{1}{n+1} x^{n+1} + c$$

2.2   Surface Area Bounded by Curves

2.3. First, we need to find the intersection points of the curves as follows:

$$2x^2 - 2x = x^2 \Rightarrow x^2 - 2x = 0 \Rightarrow x = 0, 2$$

Then:

$$S = \int_{x_1}^{x_2} (y_2 - y_1) \, dx$$

$$\Rightarrow S = \int_0^2 (x^2 - 2x^2 + 2x) \, dx = \int_0^2 (-x^2 + 2x) \, dx$$

$$\Rightarrow S = \left(-\frac{x^3}{3} + x^2 \right) \Big|_0^2 = \left(-\frac{2^3}{3} + 2^2 \right) - (0 + 0)$$

$$\Rightarrow S = \frac{4}{3}$$

Choice (3) is the answer.

In this problem, the rule below was used.

$$\int x^n dx = \frac{1}{n+1} x^{n+1} + c$$

2.4. The problem can be solved as follows.

$$S = \int_{x_1}^{x_2} (y_2 - y_1) dx$$

$$\Rightarrow S = \int_{\frac{\pi}{4}}^{\frac{5\pi}{4}} (\sin x - \cos x) dx$$

$$\Rightarrow S = (-\cos x - \sin x) \Big|_{\frac{\pi}{4}}^{\frac{5\pi}{4}} = \left(-\cos \frac{5\pi}{4} - \sin \frac{5\pi}{4}\right) - \left(-\cos \frac{\pi}{4} - \sin \frac{\pi}{4}\right) = \left(\frac{\sqrt{2}}{2} + \frac{\sqrt{2}}{2}\right) - \left(-\frac{\sqrt{2}}{2} - \frac{\sqrt{2}}{2}\right)$$

$$\Rightarrow S = 2\sqrt{2}$$

Choice (3) is the answer.

In this problem, the rules below were used.

$$\int \sin x dx = -\cos x + c$$

$$\int \cos x dx = \sin x + c$$

$$\cos \frac{5\pi}{4} = -\frac{\sqrt{2}}{2}$$

$$\sin \frac{5\pi}{4} = -\frac{\sqrt{2}}{2}$$

$$\cos \frac{\pi}{4} = \frac{\sqrt{2}}{2}$$

$$\sin \frac{\pi}{4} = \frac{\sqrt{2}}{2}$$

2.5. First, we need to find the intersection points of the curves as follows:

$$\begin{cases} y_1 = x^2 \\ y_2 = \sqrt{x} \end{cases}$$

$$\Rightarrow y_2 = y_1 \Rightarrow \sqrt{x} = x^2 \Rightarrow \sqrt{x}(x\sqrt{x} - 1) = 0 \Rightarrow x = 0, 1$$

$$S = \int_{x_1}^{x_2} (y_2 - y_1)dx$$

$$\Rightarrow S = \int_0^1 \left(\sqrt{x} - x^2\right)dx$$

$$\Rightarrow S = \left(\frac{2}{3}x^{\frac{3}{2}} - \frac{x^3}{3}\right)\Big|_0^1 = \frac{2}{3} - \frac{1}{3}$$

$$\Rightarrow S = \frac{1}{3}$$

Choice (3) is the answer.

In this problem, the rule below was used.

$$\int x^n dx = \frac{1}{n+1}x^{n+1} + c$$

2.6. First, we need to find the intersection points of the curves as follows:

$$y_2 = y_1 \Rightarrow x^3 + 2x^2 + x = 0 \Rightarrow x(x^2 + 2x + 1) = x(x+1)^2 = 0 \Rightarrow x = 0, -1, -1$$

$$S = \int_{x_1}^{x_2} (y_2 - y_1)dx$$

$$\Rightarrow S = \int_{-1}^0 (x^3 + 2x^2 + x)dx$$

$$\Rightarrow S = \left(\frac{x^4}{4} + \frac{2}{3}x^3 + \frac{x^2}{2}\right)\Big|_{-1}^0 = 0 - \left(\frac{1}{4} - \frac{2}{3} + \frac{1}{2}\right) = -\left(\frac{3-8+6}{12}\right) = -\frac{1}{12}$$

The surface area must be a positive quantity. Therefore,

$$S = \frac{1}{12}$$

Choice (1) is the answer.

In this problem, the rule below was used.

$$\int x^n dx = \frac{1}{n+1}x^{n+1} + c$$

2.7. First, we need to find the intersection points of the curves as follows:

$$\begin{cases} y_1 = x^2 + 1 \\ y_2 = 2 \end{cases}$$

$$\Rightarrow x^2 + 1 = 2 \Rightarrow x^2 = 1 \Rightarrow x = \pm 1$$

Then:

$$S = \int_{x_1}^{x_2} (y_2 - y_1) dx$$

$$\Rightarrow S = \int_{-1}^{1} (2 - (x^2 + 1)) dx = 2 \int_0^1 (1 - x^2) dx$$

$$\Rightarrow S = 2 \left(x - \frac{x^3}{3} \right) \Big|_0^1 = 2 \left(1 - \frac{1}{3} \right) - 0$$

$$\Rightarrow S = \frac{4}{3}$$

Choice (4) is the answer.

In this problem, the rules below were used.

$$\int x^n dx = \frac{1}{n+1} x^{n+1} + c$$

$$\int_{-a}^{a} f(x) dx = 2 \int_0^a f(x) dx, \quad \text{if } f(x) \text{ is an even function}$$

2.8. As can be noticed from the function of $y^2 = x^2 - x^4$, the function is even with respect to both x and y. Therefore, we can calculate the surface area bounded in the first quadrant and then multiply its value by 4.

To determine the range of x, we need to solve the equation of $y = 0$ or $\sqrt{x^2 - x^4} = x\sqrt{1 - x^2} = 0$ that for the first quadrant gives $x = 0, 1$.

$$S = \int_{x_1}^{x_2} (y_2 - y_1) dx$$

$$\Rightarrow S = 4 \int_0^1 x\sqrt{1 - x^2} dx$$

By defining a new variable, we have:

$$1 - x^2 = u \Rightarrow -2x dx = du \Rightarrow x dx = -\frac{1}{2} du$$

$$\Rightarrow S = 4 \int -\frac{1}{2} u^{\frac{1}{2}} du$$

$$\Rightarrow S = 4 \left[-\frac{1}{3} u^{\frac{3}{2}} \right] = \left[-\frac{4}{3} (1 - x^2)^{\frac{3}{2}} \right]_0^1 = -\frac{4}{3} (0 - 1)$$

$$\Rightarrow S = \frac{4}{3}$$

Choice (4) is the answer.

2.9. The problem can be solved as follows.

$$S = \int (y_1(x) - y_2(x))dx$$

$$\Rightarrow S = 2\int_0^1 \left(4 - \frac{4}{1+x^2}\right)dx$$

$$\Rightarrow S = 2(4x - 4\arctan x)\Big|_0^1 = 2\left(4 - 4\left(\frac{\pi}{4}\right) - 0\right) = 2(4 - \pi)$$

$$\Rightarrow S = 8 - 2\pi$$

Choice (1) is the answer.

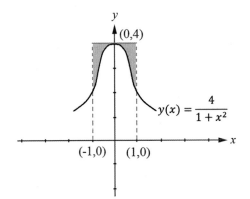

Figure 2.1 The graph of problem 2.9

In this problem, the rules below were used.

$$\int a\,dx = ax + c$$

$$\int \frac{1}{1+x^2}dx = \arctan x + c$$

$$\int_{-a}^{a} f(x)dx = 2\int_0^a f(x)dx , \qquad \text{if } f(x) \text{ is an even function}$$

$$\arctan 1 = \frac{\pi}{4}$$

$$\arctan 0 = 0$$

2.10. The problem can be solved as follows.

$$S = \int (y_1(x) - y_2(x))dx$$

$$\Rightarrow S = \int_1^\infty \left(\frac{6}{(2x+1)(x+2)} - 0 \right) dx = \int_1^\infty \left(\frac{4}{2x+1} - \frac{2}{x+2} \right) dx$$

$$\Rightarrow S = (2\ln(2x+1) - 2\ln(x+2)) \Big|_1^\infty = 2\ln \frac{2x+1}{x+2} \Big|_1^\infty = 2[\ln 2 - \ln 1]$$

$$\Rightarrow S = 2\ln 2$$

Choice (3) is the answer.

In this problem, the rules below were used.

$$\int \frac{1}{u} du = \ln u + c$$

$$\ln 1 = 0$$

2.11.   The problem can be solved as follows.

$$S = \int (y_1(x) - y_2(x)) dx$$

$$\Rightarrow S = \int_{-\sqrt{2}}^{\sqrt{2}} \frac{1}{2+x^2} dx = 2 \int_0^{\sqrt{2}} \frac{1}{2+x^2} dx = 2 \int_0^{\sqrt{2}} \frac{1}{2\left(1 + \left(\frac{x}{\sqrt{2}}\right)^2\right)} dx$$

$$\Rightarrow S = \left(\sqrt{2} \, \text{arc} \left(\tan \left(\frac{x}{\sqrt{2}} \right) \right) \right) \Big|_0^{\sqrt{2}} = \sqrt{2} \left(\frac{\pi}{4} - 0 \right)$$

$$\Rightarrow S = \frac{\pi\sqrt{2}}{4}$$

Choice (1) is the answer.

In this problem, the rules below were used.

$$\int \frac{1}{\left(1 + \left(\frac{x}{a}\right)^2\right)} dx = a \, \text{arc} \left(\tan \left(\frac{x}{a} \right) \right) + c$$

$$\int_{-a}^{a} f(x) dx = 2 \int_0^a f(x) dx, \quad \text{if } f(x) \text{ is an even function}$$

$$\text{arc} \tan 1 = \frac{\pi}{4}$$

$$\text{arc} \tan 0 = 0$$

2.12. From list of integral of functions or by using the method of integration by parts, we know that:

$$\int \ln u\, du = u \ln u - u + c$$

The problem can be solved by defining a new variable as follows.

$$x = t^2 \Rightarrow \sqrt{x} = t \Rightarrow \frac{dx}{\sqrt{x}} = 2dt$$

$$S = \int (y_1(x) - y_2(x))dx$$

$$\Rightarrow S = \int_1^{e^2} \left(\frac{\ln x}{\sqrt{x}} - 0\right)dx$$

$$\Rightarrow S = \int_{t_1}^{t_2} \ln t^2 (2dt) = 4\int_{t_1}^{t_2} \ln t\, dt = 4(t\ln t - t)\Big|_{t_1}^{t_2}$$

$$\Rightarrow S = 4\left(\sqrt{x}\ln(\sqrt{x}) - \sqrt{x}\right)\Big|_1^{e^2}$$

$$\Rightarrow S = 4[(e\ln e - e) - (1\ln 1 - 1)]$$

$$\Rightarrow S = 4$$

Choice (2) is the answer.

In this problem, the rules below were used.

$$\ln a^b = b \ln a$$

$$\ln e = 1$$

$$\ln 1 = 0$$

2.13. From list of integral of functions, we know that:

$$\int a^x dx = \frac{a^x}{\ln a} + c$$

$$\int x^n dx = \frac{1}{n+1}x^{n+1} + c$$

The problem can be solved as follows.

$$S = \int (y_1(x) - y_2(x))dx$$

$$\Rightarrow S = \int_0^1 \left(2^x - \left(1 - x^2\right)\right)dx = \int_0^1 \left(2^x - 1 + x^2\right)dx$$

$$\Rightarrow S = \left(\frac{2^x}{\ln 2} - x + \frac{x^3}{3}\right)\Bigg|_0^1 = \left(\frac{2}{\ln 2} - 1 + \frac{1}{3}\right) - \left(\frac{1}{\ln 2}\right)$$

$$\Rightarrow S = \frac{1}{\ln 2} - \frac{2}{3}$$

Choice (3) is the answer.

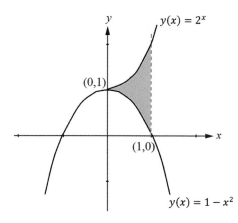

Figure 2.2 The graph of problem 2.13

2.14. The problem can be solved as follows.

$$S = \int \left(y_1(x) - y_2(x)\right)dx$$

$$\Rightarrow S = \int_0^1 \left(\frac{2}{1 + x^2} - x\right)dx$$

$$\Rightarrow S = \left(2\arctan x - \frac{x^2}{2}\right)\Bigg|_0^1 = \left(2\arctan 1 - \frac{1}{2}\right) - \left(2\arctan 0 - 0\right)$$

$$\Rightarrow S = \frac{\pi}{2} - \frac{1}{2}$$

Choice (3) is the answer.

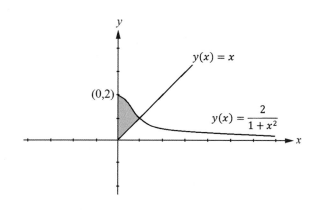

Figure 2.3 The graph of problem 2.14

In this problem, the rules below were used.

$$\int \frac{1}{1+x^2} dx = \arctan x + c$$

$$\int x^n dx = \frac{1}{n+1} x^{n+1} + c$$

$$\arctan 1 = \frac{\pi}{4}$$

$$\arctan 0 = 0$$

2.15. First, we need to determine the points of intersection of two graphs related to the first surface area, that is, $y_1(x) = \cos x$ and $y_2(x) = \cos (x - c)$.

$$y_1(x) = y_2(x) \Rightarrow \cos(x - c) = \cos x$$

$$\Rightarrow (x - c) = \pm x \Rightarrow x - c = -x \Rightarrow x = \frac{c}{2}$$

Also, from the problem, we have $x = 0$. Thus, the problem can be solved as follows.

$$S_1 = \int (y_1(x) - y_2(x)) dx$$

$$\Rightarrow S_1 = \int_0^{\frac{c}{2}} [\cos x - \cos(x - c)] dx$$

$$\Rightarrow S_1 = [\sin x - \sin(x - c)]_0^{\frac{c}{2}} = \sin\left(\frac{c}{2}\right) - \sin\left(-\frac{c}{2}\right) - (0 - \sin(-c)) = 2 \sin \frac{c}{2} - \sin c$$

Likewise, we need to determine the points of intersection of two graphs related to the second surface area, that is, $y_3(x) = 0$ and $y_4 = \cos (x - c)$.

$$y_3(x) = y_4(x) \Rightarrow \cos(x - c) = 0$$

$$\Rightarrow x - c = \frac{\pi}{2} \Rightarrow x = c + \frac{\pi}{2}$$

Also, from the problem, we have $x = \pi$. Therefore:

$$S_2 = \int (y_3(x) - y_4(x)) dx$$

$$S_2 = \int_{c+\frac{\pi}{2}}^{\pi} [0 - \cos(x - c)] dx$$

$$\Rightarrow S_2 = -[\sin(x - c)]_{c+\frac{\pi}{2}}^{\pi} = -\sin(\pi - c) + \sin\left(\frac{\pi}{2}\right) = -\sin c + 1$$

Based on the problem, we have:

$$S_1 = S_2$$

$$\Rightarrow 2\sin\frac{c}{2} - \sin c = 1 - \sin c \Rightarrow \sin\frac{c}{2} = \frac{1}{2} \Rightarrow \frac{c}{2} = \frac{\pi}{6}$$

$$\Rightarrow c = \frac{\pi}{3}$$

Choice (2) is the answer.

In this problem, the rules below were used.

$$\int \cos x = \sin x + c$$

$$\int \cos(x + a) = \sin(x + a) + c$$

$$\sin(\pi - c) = \sin(c) \quad \text{for } 0 < c < \frac{\pi}{2}$$

$$\sin\left(\frac{\pi}{2}\right) = 1$$

2.15. The problem can be solved by using the method of integration by parts or using the list of integral of functions as follows.

From list of integral of functions, we know that:

$$\int x^n \ln x dx = \frac{x^{n+1}}{n+1}\left(\ln x - \frac{1}{n+1}\right) + c \ , \quad n \neq -1$$

$$S = \int (y_1(x) - y_2(x))dx$$

$$\Rightarrow S = \int_1^2 \left(x\ln\sqrt{x} - 0\right)dx = \frac{1}{2}\int_1^2 x\ln x dx$$

$$\Rightarrow S = \frac{1}{2}\frac{x^2}{2}\left(\ln x - \frac{1}{2}\right)\Big|_1^2 = \left(\ln 2 - \frac{1}{2}\right) - \frac{1}{4}\left(0 - \frac{1}{2}\right)$$

$$\Rightarrow S = \ln 2 - \frac{3}{8}$$

Choice (4) is the answer.

In this problem, the rules below were used.

$$\ln a^b = b\ln a$$

$$\ln 1 = 0$$

2.17. From list of integral of functions, we know that:

$$\int \frac{1}{u} du = \ln u + c$$

$$\int x^n dx = \frac{1}{n+1} x^{n+1} + c$$

Also, from trigonometry, we know that:

$$\tanh x = \frac{\sinh x}{\cosh x}$$

The problem can be solved as follows.

$$S = \int (y_1(x) - y_2(x)) dx$$

$$\Rightarrow S = \int_0^{+\infty} (1 - \tanh x) dx = \int_0^{+\infty} \left[1 - \frac{\sinh x}{\cosh x} \right] dx$$

$$\Rightarrow S = [x - \ln(\cosh x)]_0^{+\infty} = (x - \ln(\cosh x)) \Big|_{x=\infty} - (x - \ln(\cosh x)) \Big|_{x=0}$$

As we know, $\cosh x \sim \frac{e^x}{2}$ when $x \to +\infty$. Therefore, $x - \ln(\cosh x) \sim x - \ln \frac{e^x}{2}$, when $x \to +\infty$. Hence:

$$\Rightarrow S = \left(x - \ln \frac{e^x}{2} \right) \Big|_{x=\infty} - (x - \ln(\cosh x)) \Big|_{x=0}$$

$$\Rightarrow S = (x - (\ln e^x - \ln 2)) \Big|_{x=\infty} - (x - \ln(\cosh x)) \Big|_{x=0}$$

$$\Rightarrow S = (\ln 2) \Big|_{x=\infty} - (x - \ln(\cosh x)) \Big|_{x=0}$$

$$\Rightarrow S = \ln 2 - (0 - \ln 1)$$

$$\Rightarrow S = \ln 2$$

Choice (2) is the answer.

In this problem, the rules below were used.

$$\int \tanh x \, dx = \ln(\cosh x)$$

$$\ln \frac{a}{b} = \ln a - \ln b$$

$$\ln e^a = a$$

$$\cosh 0 = 1$$

$$\ln 1 = 0$$

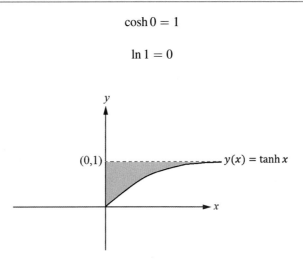

Figure 2.4 The graph of problem 2.17

2.18. Since the functions can be easily written in the form of $x = f(y)$, the formula below should be used.

$$S = \int (x_1(y) - x_2(y))dy$$

First, we need to determine the points of intersection of the two graphs of $x_1(y) = y + 4$ and $x_2(y) = \frac{1}{2}y^2$.

$$x_1(x) = x_2(x) \Rightarrow y + 4 = \frac{1}{2}y^2$$

$$\Rightarrow y^2 - 2y - 8 = (y - 4)(y + 2) = 0 \Rightarrow y = -2, 4$$

$$S = \int_{-2}^{4} \left[(y + 4) - \frac{1}{2}y^2\right] dy$$

$$\Rightarrow S = \left[\frac{1}{2}y^2 + 4y - \frac{1}{6}y^3\right]_{-2}^{4}$$

$$\Rightarrow S = \left(\frac{1}{2} \times 4^2 + 4 \times 4 - \frac{1}{6} \times 4^3\right) - \left(\frac{1}{2} \times (-2)^2 + 4 \times (-2) - \frac{1}{6} \times (-2)^3\right)$$

$$\Rightarrow S = 18$$

Choice (4) is the answer.

In this problem, the rule below was used.

$$\int x^n dx = \frac{1}{n+1}x^{n+1} + c$$

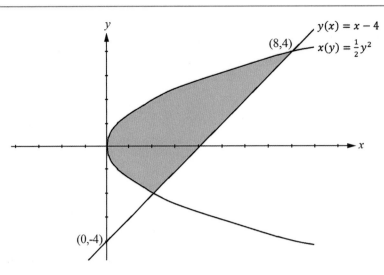

Figure 2.5 The graph of problem 2.18

2.3   Volume Resulted from Rotation of an Enclosed Region

2.19. The volume resulted from the rotation of a surface area around x-axis, enclosed between the curve of $f(x)$ and x-axis is calculated as follows:

$$V = \pi \int_{x_1}^{x_2} (f(x))^2 dx$$

In addition, from list of integral of functions, we know that:

$$\int x^n dx = \frac{1}{n+1} x^{n+1} + c$$

Therefore, based on the given information, we have:

$$V = \pi \int_{-1}^{1} \left(\sqrt{x^2 - x^3} \right)^2 dx = \pi \int_{-1}^{1} (x^2 - x^3) dx$$

$$V = \pi \left(\frac{x^3}{3} - \frac{x^4}{4} \right) \Big|_{-1}^{1} = \pi \left(\frac{1^3}{3} - \frac{1^4}{4} \right) - \pi \left(\frac{(-1)^3}{3} - \frac{(-1)^4}{4} \right)$$

$$\Rightarrow V = \frac{2\pi}{3}$$

Choice (2) is the answer.

2.20. The volume resulted from the rotation of a surface area around x-axis, enclosed between the curve of $f(x)$ and x-axis is calculated as follows:

$$V = \pi \int_{x_1}^{x_2} (f(x))^2 dx$$

Therefore,

$$V = \pi \int_0^{2\pi} \sin^2(x)dx = \pi \int_0^{2\pi} \left(\frac{1}{2} - \frac{\cos(2x)}{2}\right)dx$$

$$\Rightarrow V = \pi\left(\frac{x}{2} - \frac{1}{4}\sin(2x)\right)\Big|_0^{2\pi} = \pi(\pi - 0)$$

$$\Rightarrow V = \pi^2$$

Choice (1) is the answer.

In this problem, the rules below were used.

$$1 - \cos(2x) = 2\sin^2(x)$$

$$\int x^n dx = \frac{1}{n+1}x^{n+1} + c$$

$$\int \cos(ax)dx = \frac{1}{a}\sin(ax) + c$$

$$\sin(4\pi) = 0$$

$$\sin(0) = 0$$

2.21. The volume resulted from the rotation of a surface area around x-axis, enclosed between the curve of $f(x)$ and x-axis is calculated as follows:

$$V = \pi \int_{x_1}^{x_2} (f(x))^2 dx$$

In addition, from list of integral of functions, we know that:

$$\int \cos(ax)dx = \frac{1}{a}\sin(ax) + c$$

$$\int x^n dx = \frac{1}{n+1}x^{n+1} + c$$

Also, from trigonometry, we know that:

$$\cos^2 x = \frac{1 + \cos 2x}{2}$$

Therefore, based on the given information, we have:

$$V = \pi \int_0^{\frac{\pi}{2}} \cos^2 x dx = \pi \int_0^{\frac{\pi}{2}} \frac{1 + \cos 2x}{2}dx = \frac{\pi}{2}\int_0^{\frac{\pi}{2}}(1 + \cos 2x)dx$$

$$\Rightarrow V = \frac{\pi}{2}\left[x + \frac{1}{2}\sin 2x\right]_0^{\frac{\pi}{2}} = \frac{\pi}{2}\left[\frac{\pi}{2} + \frac{1}{2}\sin\left(2 \times \frac{\pi}{2}\right) - (0 + 0)\right]$$

$$\Rightarrow V = \frac{\pi^2}{4}$$

2.22. The volume resulted from the rotation of a surface area around x-axis, enclosed between the curve of $f(x)$ and x-axis is calculated as follows:

$$V = \pi \int_{x_1}^{x_2} (f(x))^2 dx$$

In addition, from list of integral of functions, we know that:

$$\int e^{ax} dx = \frac{e^{ax}}{a} + c$$

Therefore:

$$V = \pi \int_0^b (e^{-x})^2 dx = \pi \int_0^b e^{-2x} dx$$

$$\Rightarrow V = -\frac{\pi}{2} \left[e^{-2x}\right]_0^b = -\frac{\pi}{2} \left(e^{-2b} - 1\right)$$

$$\Rightarrow \lim_{b \to \infty} V = -\frac{\pi}{2} \left(e^{-\infty} - 1\right) = -\frac{\pi}{2} \left(0 - 1\right)$$

$$\Rightarrow \lim_{b \to \infty} V = \frac{\pi}{2}$$

In this problem, the limit below was used.

$$\lim_{x \to \infty} e^{-x} = 0$$

2.23. The surface area restricted by the function of $y_1(x)$ and $y_2(x)$ and the given boundaries can be calculated as follows.

$$S = \int_{x_1}^{x_2} (y_1(x) - y_2(x)) dx$$

Moreover, the volume resulted from the rotation of a surface area around x-axis, enclosed between the curve of $f(x)$ and x-axis is calculated as follows:

$$V = \pi \int_{x_1}^{x_2} (f(x))^2 dx$$

Therefore:

$$S = \int_0^1 \frac{1}{\sqrt{x}} dx = \int_0^1 x^{-\frac{1}{2}} dx$$

$$\Rightarrow S = \left[2x^{\frac{1}{2}}\right]_0^1 = 2 - 0$$

$$\Rightarrow S = 2$$

$$V = \pi \int_0^1 \left(\frac{1}{\sqrt{x}}\right)^2 dx = \pi \int_0^1 \frac{dx}{x}$$

$$\Rightarrow V = \pi [\ln x]_0^1 = \pi [0 - (-\infty)]$$

$$\Rightarrow V = \infty$$

Choice (3) is the answer.

In this problem, the rules below were used.

$$\int x^n dx = \frac{1}{n+1} x^{n+1} + c$$

$$\int \frac{1}{x} dx = \ln x + c$$

$$\ln 1 = 0$$

$$\lim_{x \to 0^+} \ln x = -\infty$$

2.24. The volume resulted from the rotation of a surface area around y-axis, enclosed between the curve of $f(x)$ and x-axis is calculated as follows:

$$V = \pi \int_{y_1}^{y_2} x^2 dy$$

Since x-axis is the boundary, $y_1 = 0$. Another boundary for y can be determined as follows:

$$x = 0 \Rightarrow y_2 = 1 - \frac{1}{4} \times 0 = 1$$

Therefore,

$$V = \pi \int_0^1 x^2 dy = \pi \int_0^1 (4 - 4y) dy$$

$$\Rightarrow V = \pi \left(4y - 2y^2\right)\Big|_0^1 = \pi(4 - 2)$$

$$\Rightarrow V = 2\pi$$

Choice (2) is the answer.

In this problem, the rule below was used.

$$\int x^n dx = \frac{1}{n+1} x^{n+1} + c$$

2.25. The volume resulted from the rotation of a surface area around x-axis, enclosed between the curve of $f(x)$ and x-axis is calculated as follows:

$$V = \pi \int_{x_1}^{x_2} (f(x))^2 dx$$

Therefore,

$$V = \pi \int_{\frac{\pi}{4}}^{\frac{\pi}{2}} \frac{1}{\sin^4(x)} dx = \pi \int_{\frac{\pi}{4}}^{\frac{\pi}{2}} (1 + \cot^2(x))(1 + \cot^2(x)) dx \tag{1}$$

Now, we should change the variable of the integral as follows.

$$\cot(x) \triangleq u \Rightarrow (1 + \cot^2(x)) dx = -du \tag{2}$$

Solving (1) and (2):

$$V = -\pi \int_{u_1}^{u_2} (1 + u^2) du$$

$$\Rightarrow V = -\pi \left(u + \frac{1}{3} u^3 \right) \Big|_{u_1}^{u_2}$$

$$\Rightarrow V = -\pi \left(\cot(x) + \frac{1}{3} \cot^3(x) \right) \Big|_{\frac{\pi}{4}}^{\frac{\pi}{2}}$$

$$V = -\pi \left((0 + 0) - \left(1 + \frac{1}{3} \right) \right)$$

$$\Rightarrow V = \frac{4}{3} \pi$$

Choice (4) is the answer.

In this problem, the rules below were used.

$$1 + \cot^2(x) = \frac{1}{\sin^2(x)}$$

$$\int u^n du = \frac{1}{n+1} u^{n+1} + c$$

$$\int (1 + \cot^2(x)) dx = -\cot(x) + c$$

$$\cot\left(\frac{\pi}{2} \right) = 0$$

$$\cot\left(\frac{\pi}{4}\right) = 1$$

2.26. The volume resulted from the rotation of a surface area around x-axis, enclosed between the curve of $f(x)$ and x-axis is calculated as follows:

$$V = \pi \int_{x_1}^{x_2} (f(x))^2 dx$$

Therefore:

$$V = \pi \int_{\frac{\pi}{6}}^{\frac{\pi}{2}} \left(\frac{\sqrt{\cos^3 x}}{\sin^2 x}\right)^2 = \pi \int_{\frac{\pi}{6}}^{\frac{\pi}{2}} \frac{\cos^3 x}{\sin^4 x} = \pi \int_{\frac{\pi}{6}}^{\frac{\pi}{2}} \frac{\cos x \cos^2 x}{\sin^4 x}$$

$$\Rightarrow V = \pi \int_{\frac{\pi}{6}}^{\frac{\pi}{2}} \frac{\cos x (1 - \sin^2 x)}{\sin^4 x} dx = \pi \int_{\frac{\pi}{6}}^{\frac{\pi}{2}} \left(\frac{\cos x}{\sin^4 x} - \frac{\cos x}{\sin^2 x}\right) dx$$

$$\Rightarrow V = \pi \left[\frac{-1}{3\sin^3 x} + \frac{1}{\sin x}\right]_{\frac{\pi}{6}}^{\frac{\pi}{2}}$$

$$\Rightarrow V = \pi \left(\left[\frac{-1}{3\sin^3 \frac{\pi}{2}} + \frac{1}{\sin \frac{\pi}{2}}\right] - \left[\frac{-1}{3\sin^3 \frac{\pi}{6}} + \frac{1}{\sin \frac{\pi}{6}}\right]\right) = \pi \left(\left[\frac{-1}{3} + 1\right] - \left[-\frac{8}{3} + 2\right]\right)$$

$$\Rightarrow V = \frac{4\pi}{3}$$

Choice (3) is the answer.

In this problem, the rules below were used.

$$\int u^n du = \frac{1}{n+1} u^{n+1} + c$$

$$\cos^2 x = 1 - \sin^2 x$$

$$\sin \frac{\pi}{2} = 1$$

$$\sin \frac{\pi}{6} = \frac{1}{2}$$

2.27. The volume resulted from the rotation of a surface area around x-axis, enclosed between the curve of $f(x)$ and x-axis is calculated as follows:

$$V = \pi \int_{x_1}^{x_2} (f(x))^2 dx$$

In addition, from the method of integration by parts, we know that:

$$\int u dv = uv - \int v du$$

Therefore:

$$V = \pi \int_0^1 (xe^x)^2 dx = \pi \int_0^1 x^2 e^{2x} dx$$

By applying the method of integration by parts twice, we have:

$$\Rightarrow V = \pi \left[\frac{1}{2} x^2 e^{2x} - \frac{2}{4} x e^{2x} + \frac{2}{8} e^{2x} \right]_0^1$$

$$\Rightarrow V = \pi \left(\left(\frac{1}{2} e^2 - \frac{1}{2} e^2 + \frac{1}{4} e^2 \right) - \left(0 - 0 + \frac{1}{4} \right) \right)$$

$$\Rightarrow V = \frac{\pi}{4} \left(e^2 - 1 \right)$$

Choice (3) is the answer.

2.28. The volume resulted from the rotation of a surface area around x-axis, enclosed between the curve of $f(x)$ and x-axis is calculated as follows:

$$V = \pi \int_{x_1}^{x_2} (f(x))^2 dx$$

First, we need to arrange the relation based on $y(x)$, as follows.

$$y^{\frac{1}{2}}(x) = a^{\frac{1}{2}} - x^{\frac{1}{2}} \Rightarrow y(x) = a + x - 2a^{\frac{1}{2}}x^{\frac{1}{2}}$$

Moreover, the area is restricted by y-axis; therefore, $x_1 = 0$. The other boundary of the restricted area needs to be determined as follows.

$$y(x) = 0 \Rightarrow a^{\frac{1}{2}} - x^{\frac{1}{2}} = 0 \Rightarrow x_2 = a$$

Therefore:

$$V = \pi \int_0^a \left(a + x - 2a^{\frac{1}{2}}x^{\frac{1}{2}} \right)^2 dx$$

$$\Rightarrow V = \pi \int_0^a \left(a^2 + x^2 + 4ax + 2ax - 4a^{\frac{3}{2}}x^{\frac{1}{2}} - 4a^{\frac{1}{2}}x^{\frac{3}{2}} \right) dx$$

$$\Rightarrow V = \pi \int_0^a \left(a^2 + x^2 + 6ax - 4a^{\frac{3}{2}}x^{\frac{1}{2}} - 4a^{\frac{1}{2}}x^{\frac{3}{2}} \right) dx$$

$$\Rightarrow V = \pi \left(a^2 x + \frac{x^3}{3} + 3ax^2 - 4a^{\frac{3}{2}} \times \frac{2}{3} x^{\frac{3}{2}} - 4a^{\frac{1}{2}} \times \frac{2}{5} x^{\frac{5}{2}} \right) \Big|_0^a$$

$$\Rightarrow V = \pi\left(a^3 + \frac{a^3}{3} + 3a^3 - 4a^{\frac{3}{2}} \times \frac{2}{3}a^{\frac{3}{2}} - 4a^{\frac{1}{2}} \times \frac{2}{5}a^{\frac{5}{2}}\right) - (0) = \pi\left(1 + \frac{1}{3} + 3 - \frac{8}{3} - \frac{8}{5}\right)a^3$$

$$\Rightarrow V = \frac{1}{15}\pi a^3$$

Choice (4) is the answer.

In this problem, the rule of factorization of polynomials was used as follows.

$$(a + b + c)^2 = a^2 + b^2 + c^2 + 2ab + 2bc + 2ac$$

In addition, the integral below was applied.

$$\int x^n dx = \frac{1}{n+1}x^{n+1} + c$$

2.29. The volume resulted from the rotation of a surface area around x-axis, enclosed between the curve of $f(x)$ and x-axis is calculated as follows:

$$V = \pi \int_{x_1}^{x_2} (f(x))^2 dx \tag{1}$$

$$\Rightarrow V = \pi \int_0^{\pi} \left(e^{-x}\sqrt{\sin x}\right)^2 dx = \pi \int_0^{\pi} e^{-2x}\sin x\, dx \tag{2}$$

Herein, let us first assume that:

$$V = \pi I \tag{3}$$

where,

$$I = \int_0^{\pi} e^{-2x}\sin x\, dx \tag{4}$$

The integral can be solved by using the method of integration by parts twice as follows.

$$\int u\, dv = uv - \int v\, du \tag{5}$$

$$u = e^{-2x} \Rightarrow du = -2e^{-2x}dx \tag{6}$$

$$dv = \sin x\, dx \Rightarrow v = -\cos x \tag{7}$$

$$\Rightarrow I = \int e^{-2x}\sin x\, dx = -e^{-2x}\cos x - 2\int e^{-2x}\cos x\, dx \tag{8}$$

$$u = e^{-2x} \Rightarrow du = -2e^{-2x} \tag{9}$$

$$dv = \cos x\, dx \Rightarrow v = \sin x \tag{10}$$

$$\Rightarrow I = -e^{-2x}\cos x - 2\left[e^{-2x}\sin x + 2\int e^{-2x}\sin x\, dx\right] \tag{11}$$

Solving (4) and (11):

$$\Rightarrow I = -e^{-2x}\cos x - 2e^{-2x}\sin x - 4I \tag{12}$$

$$\Rightarrow 5I = -e^{-2x}\cos x - 2e^{-2x}\sin x \tag{13}$$

$$\Rightarrow I = -\frac{e^{-2x}}{5}(\cos x + 2\sin x) \tag{14}$$

Solving (3) and (14):

$$\Rightarrow V = -\frac{\pi}{5}\left[e^{-2x}(\cos x + 2\sin x)\right]_0^\pi = -\frac{\pi}{5}\left(-e^{-2\pi} - 1\right) \tag{15}$$

$$\Rightarrow V = \frac{\pi}{5}\left(1 + e^{-2\pi}\right)$$

Choice (3) is the answer.

In this problem, the rules below were used.

$$\cos \pi = -1$$

$$\sin \pi = 0$$

$$\cos 0 = 1$$

$$\sin 0 = 0$$

2.30. The volume resulted from the rotation of a surface area around x-axis, enclosed between the curve of $f(x)$ and x-axis is calculated as follows:

$$V = \pi \int_{x_1}^{x_2} (f(x))^2 dx$$

As can be noticed from Fig. 2.6, the volume of shaded region can be achieved by subtracting the volume of empty space from the volume of sphere as follows:

$$V = V_2 - V_1 = \pi \int_0^2 \left(y_1^2(x) - y_2^2(x)\right) dx$$

Now, we need to determine the $y_1(x)$ and $y_2(x)$ as follows.

$$(x-1)^2 + y^2 = 1 \Rightarrow y^2 = 1 - (x-1)^2 \Rightarrow y_1(x) = \sqrt{1 - (x-1)^2} \Rightarrow y_1(x) = \sqrt{2x - x^2}$$

$$y_2(x) = 2x - x^2$$

$$\Rightarrow V = \pi \int_0^2 \left(\left(\sqrt{2x - x^2} \right)^2 - \left(2x - x^2 \right)^2 \right) dx$$

$$\Rightarrow V = \pi \int_0^2 \left(2x - x^2 - 4x^2 + 4x^3 - x^4 \right) dx = \pi \int_0^2 \left(-x^4 + 4x^3 - 5x^2 + 2x \right) dx$$

$$\Rightarrow V = \pi \left(\frac{-x^5}{5} + x^4 - \frac{5x^3}{3} + x^2 \right) \Big|_0^2$$

$$\Rightarrow V = \pi \left(\frac{-(2)^5}{5} + 2^4 - \frac{5(2)^3}{3} + 2^2 \right) - 0 = \pi \left(\frac{-96 + 240 - 200 + 60}{15} \right)$$

$$\Rightarrow V = \frac{4\pi}{15}$$

Choice (1) is the answer.

In this problem, the integral formula below was used.

$$\int x^n dx = \frac{1}{n+1} x^{n+1} + c$$

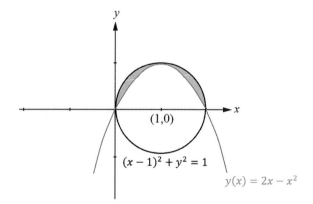

Figure 2.6 The graph of problem 1.30

2.4 Arc Length of a Curve

2.31. The arc length of a curve can be calculated as follows.

$$L = \int_a^b \sqrt{1 + \left(f'(x) \right)^2} dx$$

Therefore, first we need to determine $f'(x)$ as follows:

$$f(x) = \int_1^x \sqrt{t^4 - 1}\, dt \Rightarrow f'(x) = \sqrt{x^4 - 1}$$

Then:

$$L = \int_1^3 \sqrt{1 + \left(\sqrt{x^4 - 1}\right)^2}\, dx = \int_1^3 \sqrt{1 + x^4 - 1}\, dx = \int_1^3 x^2 dx$$

$$\Rightarrow L = \frac{1}{3}x^3 \Big|_1^3 = \frac{1}{3}(27 - 1)$$

$$\Rightarrow L = \frac{26}{3}$$

Choice (3) is the answer.

In this problem, the rules below were used.

$$F(x) = \int_{v(x)}^{u(x)} f(t)\, dt \Rightarrow F'(x) = u'(x)F(u(x)) - v'(x)F(v(x))$$

$$\int x^n dx = \frac{1}{n+1}x^{n+1} + c$$

2.32. The arc length of a parametric curve can be calculated as follows.

$$L = \int_a^b \sqrt{(x'_t)^2 + (y'_t)^2}\, dt$$

From the problem, we have:

$$\begin{cases} x(t) = e^t \cos t \\ y(t) = e^t \sin t \end{cases}$$

Thus:

$$L = \int_0^4 \sqrt{(e^t \cos t - e^t \sin t)^2 + (e^t \sin t + e^t \cos t)^2}\, dt$$

$$\Rightarrow L = \int_0^4 e^t \sqrt{\cos^2 t + \sin^2 t - 2 \sin t \cos t + \sin^2 t + \cos^2 t + 2 \sin t \cos t}\, dt$$

$$\Rightarrow L = \int_0^4 e^t \sqrt{1 + 1} = \sqrt{2} e^t \Big|_0^4$$

$$\Rightarrow L = \sqrt{2}(e^4 - 1)$$

Choice (1) is the answer.

In this problem, the rules below were used.

$$\frac{d}{dx}(u(x)v(x)) = u'(x)v(x) + v'(x)u(x)$$

$$\frac{d}{dx}e^x = e^x$$

$$\cos^2 x + \sin^2 x = 1$$

$$\int e^x = e^x + c$$

2.33. The arc length of a parametric curve can be calculated as follows.

$$L = \int_a^b \sqrt{(x_t')^2 + (y_t')^2}\, dt$$

From the problem, we have:

$$\begin{cases} x(t) = e^t(\cos t + \sin t) \\ y(t) = e^t(\cos t - \sin t) \end{cases}$$

Thus:

$$L = \int_0^1 \sqrt{(e^t(\cos t + \sin t) + e^t(-\sin t + \cos t))^2 + (e^t(\cos t - \sin t) + e^t(-\sin t - \cos t))^2}\, dt$$

$$\Rightarrow L = \int_0^1 \sqrt{(2e^t\cos t)^2 + (-2e^t \sin t)^2}\, dt = \int_0^1 \sqrt{4e^{2t}\cos^2 t + 4e^{2t}\sin^2 t}\, dt =$$

$$\Rightarrow L = \int_0^1 \sqrt{4e^{2t}(\cos^2 t + \sin^2 t)}\, dt = \int_0^1 \sqrt{4e^{2t}}\, dt = \int_0^1 2e^t\, dt$$

$$\Rightarrow L = 2e^t \Big|_0^1$$

$$\Rightarrow L = 2(e - 1)$$

Choice (4) is the answer.

In this problem, the rules below were used.

$$\frac{d}{dx}(u(x)v(x)) = u'(x)v(x) + v'(x)u(x)$$

$$\frac{d}{dx}e^x = e^x$$

$$\cos^2 x + \sin^2 x = 1$$

$$\int e^x = e^x + c$$

2.34. The arc length of a curve can be calculated as follows.

$$L = \int_a^b \sqrt{1 + (y'(x))^2}\, dx$$

Therefore, first we need to determine $f'(x)$ as follows:

$$y(x) = \frac{1}{2}x^2 - \frac{1}{4}\ln x \Rightarrow y'(x) = x - \frac{1}{4x}$$

Then:

$$L = \int_1^2 \sqrt{1 + \left(x - \frac{1}{4x}\right)^2}\, dx = \int_1^2 \sqrt{1 + x^2 - \frac{1}{2} + \frac{1}{16x^2}}\, dx = \int_1^2 \sqrt{x^2 + \frac{1}{2} + \frac{1}{16x^2}}\, dx$$

$$\Rightarrow L = \int_1^2 \sqrt{\left(x + \frac{1}{4x}\right)^2}\, dx = \int_1^2 \left(x + \frac{1}{4x}\right) dx$$

$$\Rightarrow L = \left(\frac{x^2}{2} + \frac{1}{4}\ln x\right)\Big|_1^2 = \left(\frac{2^2}{2} + \frac{1}{4}\ln 2\right) - \left(\frac{1}{2} + \frac{1}{4}\ln 1\right)$$

$$\Rightarrow L = \frac{3}{2} + \frac{1}{4}\ln 2$$

Choice (2) is the answer.

In this problem, the rules below were used.

$$\int x^n dx = \frac{1}{n+1}x^{n+1} + c$$

$$\int \frac{a}{x} dx = a \ln x + c$$

$$\frac{d}{dx}x^n = nx^{n-1}$$

$$\frac{d}{dx}\ln x = \frac{1}{x}$$

$$\ln 1 = 0$$

2.35. The arc length of a curve can be calculated as follows.

$$L = \int_a^b \sqrt{1 + (y'(x))^2}\, dx$$

Therefore, first we need to determine $f'(x)$ as follows:

$$y(x) = \frac{x^4}{8} + \frac{1}{4x^2} \Rightarrow y'(x) = \frac{x^3}{2} - \frac{1}{2x^3}$$

Then:

$$L = \int_1^2 \sqrt{1 + \left(\frac{x^3}{2} - \frac{1}{2x^3}\right)^2}\,dx = \int_1^2 \sqrt{1 + \frac{x^6}{2} - \frac{1}{2} + \frac{1}{4x^6}}\,dx$$

$$\Rightarrow L = \int_1^2 \sqrt{\frac{x^6}{2} + \frac{1}{2} + \frac{1}{4x^6}}\,dx = \int_1^2 \sqrt{\left(\frac{x^3}{2} + \frac{1}{2x^3}\right)^2}\,dx = \int_1^2 \left(\frac{x^3}{2} + \frac{1}{2x^3}\right)dx$$

$$\Rightarrow L = \left(\frac{x^4}{8} - \frac{1}{4x^2}\right)\Big|_1^2 = \left(\frac{2^4}{8} - \frac{1}{4 \times 2^2}\right) - \left(\frac{1^4}{8} - \frac{1}{4 \times 1^2}\right) = 2 - \frac{1}{16} - \frac{1}{8} + \frac{1}{4}$$

$$\Rightarrow L = \frac{33}{16}$$

Choice (4) is the answer.

In this problem, the rules below were used.

$$\int x^n dx = \frac{1}{n+1}x^{n+1} + c$$

$$\frac{d}{dx}x^n = nx^{n-1}$$

2.36. The arc length of a curve can be calculated as follows.

$$L = \int_a^b \sqrt{1 + (f'(x))^2}\,dx$$

Therefore, first we need to determine $f'(x)$ as follows:

$$f(x) = \ln(\sec x) \Rightarrow f'(x) = \tan x$$

Then,

$$L = \int_0^{\frac{\pi}{4}} \sqrt{1 + \tan^2 x}\,dx = \int_0^{\frac{\pi}{4}} \sqrt{\frac{1}{\cos^2 x}}\,dx = \int_0^{\frac{\pi}{4}} \frac{1}{\cos x}\,dx = \int_0^{\frac{\pi}{4}} \sec x\,dx$$

$$\Rightarrow L = \ln(\sec x + \tan x)\Big|_0^{\frac{\pi}{4}} = \ln\left(\sec\frac{\pi}{4} + \tan\frac{\pi}{4}\right) - \ln(\sec 0 + \tan 0) = \ln\left(\sqrt{2} + 1\right) - \ln(1 + 0)$$

$$\Rightarrow L = \ln\left(1 + \sqrt{2}\right)$$

Choice (4) is the answer.

In this problem, the rules below were used.

$$\frac{d}{dx}\ln(\sec x) = \frac{d}{dx}\ln\left(\frac{1}{\cos x}\right) = \frac{\frac{\sin x}{\cos^2 x}}{\frac{1}{\cos x}} = \frac{\sin x \cos x}{\cos^2 x} = \frac{\sin x}{\cos x} = \tan x$$

$$1 + \tan^2 x = \frac{1}{\cos^2 x} = \sec^2 x$$

$$\int \sec x\, dx = \ln(\sec x + \tan x) + c$$

$$\sec \frac{\pi}{4} = \sqrt{2}$$

$$\tan \frac{\pi}{4} = 1$$

$$\sec 0 = 1$$

$$\tan 0 = 0$$

2.37. The arc length of a curve can be calculated as follows.

$$L = \int_a^b \sqrt{1 + (f'(x))^2}\, dx$$

Therefore, first we need to determine $f'(x)$ as follows:

$$f(x) = \int_1^x \sqrt{\cosh(t)}\, dt \Rightarrow f'(x) = \sqrt{\cosh(x)}$$

Then:

$$L = \int_0^2 \sqrt{1 + \left(\sqrt{\cosh(x)}\right)^2}\, dx = \int_0^2 \sqrt{1 + \cosh(x)}\, dx$$

$$\Rightarrow L = \int_0^2 \sqrt{2\cosh^2\left(\frac{x}{2}\right)}\, dx = \int_0^2 \sqrt{2}\cosh\left(\frac{x}{2}\right) dx$$

$$\Rightarrow L = 2\sqrt{2}\sinh\left(\frac{x}{2}\right)\Big|_0^2 = 2\sqrt{2}\left[\sinh\frac{2}{2} - \sinh 0\right] = 2\sqrt{2}\sinh 1$$

$$\Rightarrow L = 2\sqrt{2}\frac{e^1 - e^{-1}}{2}$$

$$\Rightarrow L = \sqrt{2}\left(e - \frac{1}{e}\right)$$

Choice (1) is the answer.

In this problem, the rules below were used.

$$F(x) = \int_{v(x)}^{u(x)} f(t)\, dt \Rightarrow F'(x) = u'(x)F(u(x)) - v'(x)F(v(x))$$

$$1 + \cosh(x) = 2\cosh^2\left(\frac{x}{2}\right)$$

$$\int \cosh(ax) = \frac{1}{a}\sinh(ax) + c$$

$$\sinh 1 = \frac{e^1 - e^{-1}}{2}$$

$$\sinh 0 = 0$$

2.38.   The arc length of a curve can be calculated as follows.

$$L = \int_a^b \sqrt{1 + (f'(x))^2}\, dx$$

Therefore, first we need to determine $f'(x)$ as follows:

$$f(x) = \frac{1}{2}\left[x\sqrt{x^2-1} - \ln\left(x + \sqrt{x^2-1}\right)\right] \Rightarrow f'(x) = \frac{1}{2}\left[\sqrt{x^2-1} + \frac{x^2}{\sqrt{x^2-1}} - \frac{1 + \frac{x}{\sqrt{x^2-1}}}{x + \sqrt{x^2-1}}\right]$$

$$\Rightarrow f'(x) = \frac{1}{2}\left[\sqrt{x^2-1} + \frac{x^2}{\sqrt{x^2-1}} - \frac{\sqrt{x^2-1} + x}{\left(x + \sqrt{x^2-1}\right)\left(\sqrt{x^2-1}\right)}\right]$$

$$\Rightarrow f'(x) = \frac{1}{2}\left[\sqrt{x^2-1} + \frac{x^2}{\sqrt{x^2-1}} - \frac{1}{\sqrt{x^2-1}}\right]$$

$$\Rightarrow f'(x) = \frac{1}{2}\left[\frac{x^2-1 + x^2-1}{\sqrt{x^2-1}}\right] = \frac{x^2-1}{\sqrt{x^2-1}} = \sqrt{x^2-1}$$

Then:

$$L = \int_1^2 \sqrt{1 + \left(\sqrt{x^2-1}\right)^2}\, dx = \int_1^2 \sqrt{1 + (x^2-1)}\, dx = \int_1^2 x\, dx$$

$$\Rightarrow L = \frac{x^2}{2}\Big|_1^2 = \frac{2^2}{2} - \frac{1^2}{2}$$

$$\Rightarrow L = \frac{3}{2}$$

Choice (2) is the answer.

In this problem, the integral formula below was used.

$$\int x^n dx = \frac{1}{n+1}x^{n+1} + c$$

$$\frac{d}{dx}(u(x)v(x)) = u'(x)v(x) + v'(x)u(x)$$

$$\frac{d}{dx}\sqrt{u(x)} = \frac{u'(x)}{2\sqrt{u(x)}}$$

$$\frac{d}{dx}\ln u(x) = \frac{u'(x)}{u(x)}$$

2.39. The arc length of a curve can be calculated as follows.

$$L = \int_a^b \sqrt{1 + (f'(x))^2}\,dx$$

Therefore, first we need to determine $f'(x)$ as follows:

$$f(x) = \ln\left(\frac{e^x + 1}{e^x - 1}\right) = \ln(e^x + 1) - \ln(e^x - 1)$$

$$\Rightarrow f'(x) = \frac{e^x}{e^x + 1} - \frac{e^x}{e^x - 1} = \frac{e^x(e^x - 1) - e^x(e^x + 1)}{e^{2x} - 1} = \frac{-2e^x}{e^{2x} - 1}$$

Then,

$$L = \int_0^2 \sqrt{1 + \left(\frac{-2e^x}{e^{2x} - 1}\right)^2}\,dx = \int_0^2 \sqrt{1 + \frac{4e^{2x}}{(e^{2x} - 1)^2}}\,dx$$

$$\Rightarrow L = \int_0^2 \sqrt{\frac{(e^{2x} - 1)^2 + 4e^{2x}}{(e^{2x} - 1)^2}}\,dx = \int_0^2 \sqrt{\frac{e^{4x} - 2e^{2x} + 1 + 4e^{2x}}{(e^{2x} - 1)^2}}\,dx$$

$$\Rightarrow L = \int_0^2 \sqrt{\frac{e^{4x} + 2e^{2x} + 1}{(e^{2x} - 1)^2}}\,dx = \int_0^2 \sqrt{\frac{(e^{2x} + 1)^2}{(e^{2x} - 1)^2}}\,dx$$

$$\Rightarrow L = \int_0^2 \frac{e^{2x} + 1}{e^{2x} - 1}\,dx = \int_0^2 \frac{e^x + e^{-x}}{e^x - e^{-x}}\,dx$$

$$\Rightarrow L = \ln\left(e^x - e^{-x}\right)\Big|_1^2 = \ln\left(e^2 - e^{-2}\right) - \ln\left(e^1 - e^{-1}\right)$$

$$\Rightarrow L = \ln\left(\frac{e^2 - e^{-2}}{e - e^{-1}}\right) = \ln\left(\frac{(e - e^{-1})(e + e^{-1})}{(e - e^{-1})}\right) = \ln\left(e + e^{-1}\right)$$

$$\Rightarrow L = \ln\left(e + \frac{1}{e}\right)$$

Choice (2) is the answer.

In this problem, the rules below were used.

$$\frac{d}{dx}\ln u(x) = \frac{u'(x)}{u(x)}$$

$$\frac{d}{dx}e^x = e^x$$

$$\int \frac{1}{u(x)}du = \ln u(x) + c$$

2.5   Surface Area of a Solid of Revolution

2.40. If the function of $f(x)$ is rotated around x-axis, the surface area of the solid of revolution can be calculated as follows.

$$S = 2\pi \int_a^b f(x)\sqrt{1 + (f'(x))^2}dx$$

Moreover, if the function is rotated around y-axis, the surface area of the solid of revolution is calculated as follows.

$$S = 2\pi \int_a^b x\sqrt{1 + (f'(x))^2}dx$$

For this problem, we have:

$$S = 2\pi \int_0^2 \cosh x\sqrt{1 + \sinh^2 x}\,dx = 2\pi \int_0^2 \cosh x\sqrt{\cosh^2 x}\,dx$$

$$\Rightarrow S = 2\pi \int_0^2 \cosh^2 x\,dx = 2\pi \int_0^2 \frac{1}{2}(1 + \cosh 2x)\,dx$$

$$\Rightarrow S = \pi \left[x + \frac{\sinh 2x}{2}\right]_0^2 = \pi \left(2 + \frac{\sinh 4}{2}\right) - 0$$

$$\Rightarrow S = \frac{\pi}{2}(4 + \sinh 4)$$

Choice (4) is the answer.

In this problem, the rules below were used.

$$\frac{d}{dx}\cosh x = \sinh x$$

$$\sinh^2 x + \cosh^2 x = 1$$

$$1 + \cosh 2x = 2\cosh^2 x$$

$$\int x^n dx = \frac{1}{n+1}x^{n+1} + c$$

$$\int \cosh ax = \frac{\sinh ax}{a} + c$$

$$\sinh 0 = 0$$

2.41. If the function of $f(x)$ is rotated around x-axis, the surface area of the solid of revolution is calculated as follows.

$$S = 2\pi \int_a^b f(x)\sqrt{1 + (f'(x))^2}\, dx$$

Moreover, if the function is rotated around y-axis, the surface area of the solid of revolution is calculated as follows.

$$S = 2\pi \int_a^b x\sqrt{1 + (f'(x))^2}\, dx$$

For this problem, we have:

$$3y(x) - x^3 = 0 \Rightarrow y(x) = \frac{1}{3}x^3$$

$$\Rightarrow S = 2\pi \int_0^1 \frac{1}{3}x^3\sqrt{1 + (x^2)^2} = \frac{2\pi}{3}\int_0^1 x^3\sqrt{1 + x^4}\, dx$$

The integral can be solved by defining a new variable as follows.

$$1 + x^4 = u$$

$$4x^3\, dx = du \Rightarrow x^3\, dx = \frac{du}{4}$$

$$\Rightarrow S = \frac{2\pi}{3}\int u^{\frac{1}{2}}\left(\frac{du}{4}\right) = \frac{\pi}{6}\int u^{\frac{1}{2}}\, du = \frac{\pi}{6}\left[\frac{2}{3}u^{\frac{3}{2}}\right]$$

$$\Rightarrow S = \frac{\pi}{6}\left[\frac{2}{3}\left(1 + x^4\right)^{\frac{3}{2}}\right]_0^1 = \frac{\pi}{9}\left((1 + 1)^{\frac{3}{2}} - (1 + 0)^{\frac{3}{2}}\right)$$

$$\Rightarrow S = \frac{\pi}{9}\left(2\sqrt{2} - 1\right)$$

Choice (1) is the answer.

In this problem, the rule below was used.

$$\int u^n\, dx = \frac{1}{n + 1}u^{n+1} + c$$

2.42. If the function of $f(x)$ is rotated around x-axis, the surface area of the solid of revolution is calculated as follows.

$$S = 2\pi \int_a^b f(x)\sqrt{1 + (f'(x))^2}\, dx$$

Moreover, if the function is rotated around y-axis, the surface area of the solid of revolution is calculated as follows.

$$S = 2\pi \int_a^b x\sqrt{1 + (f'(x))^2}\,dx$$

From $y = x^2$ and $y \leq 2$, we have $-\sqrt{2} \leq x \leq \sqrt{2}$; however, the range of integration must be $0 \leq x \leq \sqrt{2}$. Therefore:

$$S = 2\pi \int_0^{\sqrt{2}} x\sqrt{1 + (2x)^2}\,dx = 2\pi \int_0^{\sqrt{2}} x\sqrt{1 + 4x^2}\,dx$$

The problem can be solved by defining a new variable as follows.

$$u = 1 + 4x^2 \Rightarrow du = 8xdx \Rightarrow \frac{du}{8} = xdx$$

$$S = \frac{2\pi}{8} \int_0^{\sqrt{2}} \sqrt{u}\,du = \frac{\pi}{4}\left[\frac{2}{3}u^{\frac{3}{2}}\right]$$

$$\Rightarrow S = \frac{\pi}{4}\left[\frac{2}{3}(1 + 4x^2)^{\frac{3}{2}}\right]_0^{\sqrt{2}} = \frac{\pi}{6}\left[\left(1 + 4 \times \left(\sqrt{2}\right)^2\right)^{\frac{3}{2}} - 1\right] = \frac{\pi}{6}\left(9^{\frac{3}{2}} - 1\right) = \frac{\pi}{6}(27 - 1) = \frac{26\pi}{6}$$

$$S = \frac{13\pi}{3}$$

Choice (2) is the answer.

In this problem, the rule below was used.

$$\int u^n\,du = \frac{1}{n+1}u^{n+1} + c$$

2.42. The volume resulted from the rotation of a surface area around x-axis, enclosed between the curve of $f(x)$ and x-axis is calculated as follows:

$$V = \pi \int_{x_1}^{x_2} (f(x))^2\,dx$$

In addition, if the function of $f(x)$ is rotated around x-axis, the surface area of the solid of revolution is calculated as follows.

$$S = 2\pi \int_a^b f(x)\sqrt{1 + (f'(x))^2}\,dx$$

Moreover, if the function is rotated around y-axis, the surface area of the solid of revolution is calculated as follows.

$$S = 2\pi \int_a^b x\sqrt{1 + (f'(x))^2}\,dx$$

For this problem, we have:

$$V = \pi \int_0^5 \left(6\cosh\frac{x}{6}\right)^2\,dx = 36\pi \int_0^5 \cosh^2\frac{x}{6}\,dx$$

Moreover:

$$S = 2\pi \int_a^b 6\cosh\frac{x}{6}\sqrt{1+\left(\sinh\frac{x}{6}\right)^2}\,dx = 2\pi \int_a^b 6\cosh\frac{x}{6}\sqrt{\cosh^2\frac{x}{6}}\,dx$$

$$\Rightarrow S = 12\pi \int_a^b 6\cosh^2\frac{x}{6}\,dx$$

Therefore:

$$\frac{V}{S} = \frac{36\pi \displaystyle\int_0^5 \cosh^2\frac{x}{6}\,dx}{12\pi \displaystyle\int_0^5 \cosh^2\frac{x}{6}\,dx}$$

$$\Rightarrow \frac{V}{S} = 3$$

Choice (1) is the answer.

In this problem, the rule below was used.

$$\cosh^2 x - \sinh^2 x = 1$$

2.43. If the function of $f(x)$ is rotated around x-axis, the surface area of the solid of revolution is calculated as follows.

$$S = 2\pi \int_a^b f(x)\sqrt{1+(f'(x))^2}\,dx$$

Moreover, if the function is rotated around y-axis, the surface area of the solid of revolution is calculated as follows.

$$S = 2\pi \int_a^b x\sqrt{1+(f'(x))^2}\,dx$$

Therefore:

$$S = 2\pi \int_0^{\sqrt{\ln(1396)}} x\sqrt{1+\sinh^2(x^2)}\,dx = 2\pi \int_0^{\sqrt{\ln(1396)}} x\cosh(x^2)\,dx$$

$$\Rightarrow S = 2\pi \left[\frac{1}{2}\sinh(x^2)\right]_0^{\sqrt{\ln(1396)}} = \pi\sinh(\ln(1396)) - 0 = \pi\sinh(\ln(1396))$$

$$\Rightarrow S = \pi\left(\frac{e^{\ln 1396} - e^{-\ln 1396}}{2}\right)$$

$$\Rightarrow S = \frac{\pi}{2}\left(1396 - \frac{1}{1396}\right)$$

Choice (4) is the answer.

In this problem, the rules below were used.

$$F(x) = \int_{v(x)}^{u(x)} f(t)\, dt \Rightarrow F'(x) = u'(x)F(u(x)) - v'(x)F(v(x))$$

$$\cosh^2 x - \sinh^2 x = 1$$

$$\int \cosh(u(x))du = \sinh(u(x)) + c$$

$$\sinh(0) = 0$$

$$\sinh u = \frac{e^u - e^{-u}}{2}$$

$$e^{\ln a} = a$$

$$e^{-\ln a} = \frac{1}{a}$$

2.6   Center of Gravity

2.44. The center of gravity of a flat surface from y-axis which is restricted by the functions of $y_1(x)$ and $y_2(x)$ can be calculated as follows.

$$\bar{x} = \frac{\int_a^b x(y_1(x) - y_2(x))dx}{S}$$

where, S is the whole surface area.

$$S = \int_a^b (y_1(x) - y_2(x))dx$$

In addition, the center of gravity of a flat surface from x-axis which is restricted by the functions of $y_1(x)$ and $y_2(x)$ can be calculated as follows.

$$\bar{y} = \frac{\frac{1}{2}\int_a^b (y_1(x) - y_2(x))^2 dx}{S}$$

Therefore, for this problem, we have:

$$y(x) = 0 \Rightarrow 1 - x^2 = 0 \Rightarrow x = \pm 1$$

$$\bar{y} = \frac{\frac{1}{2}\int_{-1}^{1}(1-x^2)^2dx}{\int_{-1}^{1}(1-x^2)dx} = \frac{\int_{0}^{1}(1-2x^2+x^4)dx}{2\int_{0}^{1}(1-x^2)dx}$$

$$\Rightarrow \bar{y} = \frac{\left[x-\frac{2}{3}x^3+\frac{1}{5}x^5\right]_0^1}{2\left[x-\frac{1}{3}x^3\right]_0^1} = \frac{\left(1-\frac{2}{3}+\frac{1}{5}\right)-0}{2\left(1-\frac{1}{3}\right)} = \frac{\frac{15-10+3}{15}}{\frac{4}{3}} = \frac{\frac{8}{15}}{\frac{4}{3}}$$

$$\Rightarrow \bar{y} = \frac{2}{5}$$

Choice (4) is the answer.

In this problem, the rules below were used.

$$\int_{-a}^{a} f(x)dx = 2\int_{0}^{a} f(x)dx \ , \quad \text{if } f(x) \text{ is an even function}$$

$$\int x^n dx = \frac{1}{n+1}x^{n+1} + c$$

2.46. The center of gravity of a flat surface from y-axis which is restricted by the functions of $y_1(x)$ and $y_2(x)$ can be calculated as follows.

$$\bar{x} = \frac{\int_{a}^{b} x(y_1(x)-y_2(x))dx}{S}$$

where, S is the whole surface area.

$$S = \int_{a}^{b}(y_1(x)-y_2(x))dx$$

In addition, the center of gravity of a flat surface from x-axis which is restricted by the functions of $y_1(x)$ and $y_2(x)$ can be calculated as follows.

$$\bar{y} = \frac{\frac{1}{2}\int_{a}^{b}(y_1(x)-y_2(x))^2dx}{S}$$

Therefore, for this problem, we have:

$$y(x) = 0 \Rightarrow 1-x^2 = 0 \Rightarrow x = \pm 1$$

$$\bar{x} = \frac{\int_{-1}^{1} x(1-x^2)dx}{\int_{-1}^{1}(1-x^2)dx}$$

The function of $x(1-x^2)$ is an odd function; therefore, its integral over a symmetric boundary is zero. In other words:

$$\frac{1}{2}\int_{-1}^{1} x\left(1 - x^2\right)dx = 0$$

Thus:

$$\Rightarrow \bar{x} = 0$$

Choice (1) is the answer.

In this problem, the rule below was used.

$$\int_{-a}^{a} f(x)dx = 0 \ , \quad \text{if } f(x) \text{ is an odd function}$$

2.47. The center of gravity of a flat curve from y-axis can be calculated as follows.

$$\bar{x} = \frac{\int_{a}^{b} x\sqrt{1 + (y'(x))^2}\,dx}{L}$$

where, L is the whole length of curve.

$$L = \int_{a}^{b} \sqrt{1 + (y'(x))^2}\,dx$$

In addition, the center of gravity of a flat curve from y-axis can be calculated as follows.

$$\bar{y} = \frac{\int_{a}^{b} y\sqrt{1 + (y'(x))^2}\,dx}{L}$$

Moreover, if the function is in the parametric form of $x = x(t)$ and $y = y(t)$, the center of gravity of a flat curve from y-axis and x-axis can be calculated as follows, respectively.

$$\bar{x} = \frac{\int_{a}^{b} x(x)\sqrt{(x'_t)^2 + (y'_t)^2}\,dt}{L}$$

$$\bar{y} = \frac{\int_{a}^{b} y(x)\sqrt{(x'_t)^2 + (y'_t)^2}\,dt}{L}$$

where, L is the whole length of curve.

$$L = \int_{a}^{b} \sqrt{(x'_t)^2 + (y'_t)^2}\,dt$$

Therefore, for this problem, we have:

$$\bar{y} = \frac{\int_0^\pi y(x)\sqrt{(x_t')^2 + (y_t')^2}\,dt}{\int_0^\pi \sqrt{(x_t')^2 + (y_t')^2}\,dt} = \frac{\int_0^\pi (1-\cos t)\sqrt{(1+\cos t)^2 + (\sin t)^2}\,dt}{\int_0^\pi \sqrt{(1+\cos t)^2 + (\sin t)^2}\,dt}$$

$$\Rightarrow \bar{y} = \frac{\int_0^\pi (1-\cos t)\sqrt{1+2\cos t + (\cos t)^2 + (\sin t)^2}\,dt}{\int_0^\pi \sqrt{1+2\cos t + (\cos t)^2 + (\sin t)^2}\,dt} = \frac{\int_0^\pi (1-\cos t)\sqrt{2(1+\cos t)}\,dt}{\int_0^\pi \sqrt{2(1+\cos t)}\,dt}$$

$$\Rightarrow \bar{y} = \frac{\int_0^\pi 2\sin^2\frac{t}{2}\sqrt{4\cos^2\frac{t}{2}}\,dt}{\int_0^\pi \sqrt{4\cos^2\frac{t}{2}}\,dt} = \frac{\int_0^\pi \left(2\sin^2\frac{t}{2}\right)\left(2\cos\frac{t}{2}\right)dt}{\int_0^\pi 2\cos\frac{t}{2}\,dt} = \frac{\int_0^\pi 4\sin^2\frac{t}{2}\cos\frac{t}{2}\,dt}{\int_0^\pi 2\cos\frac{t}{2}\,dt}$$

$$\Rightarrow \bar{y} = \frac{\frac{8}{3}\sin^3\frac{t}{2}\Big|_0^\pi}{4\sin\frac{t}{2}\Big|_0^\pi} = \frac{\frac{8}{3}\left(\sin^3\frac{\pi}{2} - \sin^3 0\right)}{4\left(\sin\frac{\pi}{2} - \sin 0\right)} = \frac{\frac{8}{3}}{4}$$

$$\Rightarrow \bar{y} = \frac{2}{3}$$

Choice (3) is the answer.

In this problem, the rules below were used.

$$(\cos t)^2 + (\sin t)^2 = 1$$

$$1 + \cos t = 2\cos^2\frac{t}{2}$$

$$1 - \cos t = 2\sin^2\frac{t}{2}$$

$$\int \cos ax\,dx = \frac{1}{a}\sin ax + c$$

$$\int u^n\,du = \frac{1}{n+1}u^{n+1} + c$$

$$\sin\frac{\pi}{2} = 1$$

$$\sin 0 = 0$$

References

1. Rahmani-Andebili, M. (2023). Calculus I (2nd Ed.) – Practice Problems, Methods, and Solutions, Springer Nature, 2021.
2. Rahmani-Andebili, M. (2021). Calculus – Practice Problems, Methods, and Solutions, Springer Nature, 2021.
3. Rahmani-Andebili, M. (2021). Precalculus – Practice Problems, Methods, and Solutions, Springer Nature, 2021.

Problems: Sequences and Series and Their Applications

3

Abstract

In this chapter, the basic and advanced problems of sequences and series as well as their applications are presented. The subjects include limit of general term of a sequences and series, convergence or divergence status of a sequences and series, convergence range and radius of series, Maclaurin expansion, Taylor expansion, concept of growth rate, concept of equivalent functions, P-series, harmonic series, telescoping series, and Stirling's approximation. In this chapter, the problems are categorized in different levels based on their difficulty levels (easy, normal, and hard) and calculation amounts (small, normal, and large). Additionally, the problems are ordered from the easiest problem with the smallest computations to the most difficult problems with the largest calculations.

3.1. Consider the sequence below and calculate its limit [1–3].

$$\left\{ \frac{2^n}{(n+2)!} \right\}$$

Difficulty level ○ Easy ● Normal ○ Hard
Calculation amount ● Small ○ Normal ○ Large
1) 0
2) ∞
3) 1
4) Not available

3.2. Consider the sequence below and calculate its limit.

$$\left\{ \sqrt[8]{\frac{n! + 2n^8 + \ln n}{n! + 5^n + 4n}} \right\}$$

Difficulty level ○ Easy ● Normal ○ Hard
Calculation amount ● Small ○ Normal ○ Large
1) 0
2) 1
3) 2
4) Not available

3.3. The general term of a sequence is as follows. Calculate the limit of the sequence.

$$a_n = \frac{\sqrt{n}}{\sqrt{4n^3 + \sin n + 1}} + \frac{\sqrt{n}}{\sqrt{4n^3 + \sin n + 2}} + \cdots + \frac{\sqrt{n}}{\sqrt{4n^3 + \sin n + n}}$$

Difficulty level　　○ Easy　● Normal　○ Hard
Calculation amount　○ Small　● Normal　○ Large

1) $\dfrac{1}{2}$

2) 1

3) ∞

4) 0

3.4. Determine the coefficient of x^3 in the Maclaurin expansion of the function of $f(x)$ for the following information.

$$f(0) = 1$$

$$f'(x) = 1 + (f(x))^{10}$$

Difficulty level　　○ Easy　● Normal　○ Hard
Calculation amount　○ Small　● Normal　○ Large

1) $\dfrac{140}{3}$

2) $\dfrac{190}{3}$

3) 280

4) 380

3.5. Some of the terms of a sequence is as follows. Calculate the limit of the sequence.

$$1, -\frac{2}{2}, \frac{3}{4}, \frac{-4}{8}, \frac{5}{16}, \cdots$$

Difficulty level　　○ Easy　○ Normal　● Hard
Calculation amount　● Small　○ Normal　○ Large

1) 0

2) $\dfrac{1}{2}$

3) $\dfrac{3}{4}$

4) $-\dfrac{1}{2}$

3.6. Determine the status of the series below.

$$S = \sum_{n=1}^{+\infty} \frac{1 + \cos n}{n^2}$$

Difficulty level　　○ Easy　○ Normal　● Hard
Calculation amount　● Small　○ Normal　○ Large

1) Convergent

2) Divergent

3) Can be convergent or divergent

4) Impossible to determine

3.7. Determine the status of the series below.

$$S = \sum_{n=1}^{+\infty} \frac{10n^2 + 9n + 8}{12n^3 + 11n^2 + 10n + 9}$$

Difficulty level   ○ Easy   ○ Normal   ● Hard

Calculation amount   ● Small   ○ Normal   ○ Large

1) Convergent

2) Divergent

3) Can be convergent or divergent

4) Impossible to determine

3.8. Calculate the final answer of the following term.

$$S = \sum_{n=1}^{\infty} \frac{\sqrt{n+1} - \sqrt{n}}{\sqrt{n^2 + n}}$$

Difficulty level   ○ Easy   ○ Normal   ● Hard

Calculation amount   ● Small   ○ Normal   ○ Large

1) $\frac{1}{2}$

2) 1

3) 2

4) ∞

3.9. Determine the convergence range of the series below.

$$S = \sum_{n=1}^{\infty} n^{1390} x^n$$

Difficulty level   ○ Easy   ○ Normal   ● Hard

Calculation amount   ● Small   ○ Normal   ○ Large

1) $[-1, 1]$

2) $(-1, 1)$

3) $(-1, 1]$

4) $[-1, 1)$

3.10. Determine the convergence radius of the series below.

$$S = \sum_{n=0}^{\infty} \frac{n! \, x^n}{3^{n^2}}$$

Difficulty level   ○ Easy   ○ Normal   ● Hard

Calculation amount   ● Small   ○ Normal   ○ Large

1) 0

2) 1

3) $2 \ln 3$

4) ∞

3.11. Determine the convergence radius of the series below.

$$S = \sum_{n=1}^{\infty} \frac{(n!)x^n}{(n+1)^n}$$

Difficulty level   ○ Easy   ○ Normal   ● Hard

Calculation amount   ● Small   ○ Normal   ○ Large

1) $\dfrac{1}{e}$

2) e

3) $\dfrac{1}{e}$

4) ∞

3.12. Determine the Taylor series of the function below around $x = 2$.

$$f(x) = \frac{1}{5-x}$$

Difficulty level   ○ Easy   ○ Normal   ● Hard

Calculation amount   ○ Small   ● Normal   ○ Large

1) $f(x) = \dfrac{1}{3} + \dfrac{1}{3^2}(x-2) + \dfrac{1}{3^3}(x-2)^2 + \cdots$

2) $f(x) = \dfrac{1}{5} + (x-2) + (x-2)^2 + \cdots$

3) $f(x) = \dfrac{1}{3} - \dfrac{1}{3^2}(x-2) + \dfrac{1}{3^3}(x-2)^3 - \cdots$

4) $f(x) = \dfrac{1}{5}\left(1 - (x-2) + (x-2)^2 + (x-2)^3 + \cdots\right)$

3.13. The general term of two sequence are as follows. What are the limits of the sequences?

$$a_n = \sqrt[3]{n^2 - n^3} + n$$

$$b_n = \frac{\sqrt{n^2+1} - \sqrt{n}}{\sqrt[4]{n^3+n} + \sqrt{n}}$$

Difficulty level   ○ Easy   ○ Normal   ● Hard

Calculation amount   ○ Small   ● Normal   ○ Large

1) $\lim\limits_{n\to\infty} a_n = 1,\ \lim\limits_{n\to\infty} b_n = 0$

2) $\lim\limits_{n\to\infty} a_n = \infty,\ \lim\limits_{n\to\infty} b_n = \infty$

3) $\lim\limits_{n\to\infty} a_n = \dfrac{1}{3},\ \lim\limits_{n\to\infty} b_n = \infty$

4) $\lim\limits_{n\to\infty} a_n = \dfrac{1}{3},\ \lim\limits_{n\to\infty} b_n = 0$

3.14. The general term of a sequence is as follows. Calculate the limit of the sequence.

$$a_n = \frac{1 + \frac{1}{2} + \frac{1}{3} + \cdots + \frac{1}{n^3}}{\ln n}$$

Difficulty level   ○ Easy   ○ Normal   ● Hard
Calculation amount   ○ Small   ● Normal   ○ Large
1) 0
2) $\frac{1}{3}$
3) 3
4) ∞

3.15. The general term of a sequence is as follows. Calculate the limit of the sequence.

$$a_n = \frac{1}{\sqrt{n^2 + 1}} + \frac{1}{\sqrt{n^2 + 2}} + \cdots \frac{1}{\sqrt{n^2 + n}}$$

Difficulty level   ○ Easy   ○ Normal   ● Hard
Calculation amount   ○ Small   ● Normal   ○ Large
1) $\ln \sqrt{2}$
2) $\frac{1}{2}$
3) 1
4) Now available

3.16. Calculate the sum of the series below.

$$S = \frac{1}{1} + \frac{1}{1+2} + \frac{1}{1+2+3} + \frac{1}{1+2+3+4} + \cdots$$

Difficulty level   ○ Easy   ○ Normal   ● Hard
Calculation amount   ○ Small   ● Normal   ○ Large
1) 2
2) $\frac{5}{2}$
3) $\frac{9}{4}$
4) 2 ln 2

3.17. Which one of the sequences is divergent?
Difficulty level   ○ Easy   ○ Normal   ● Hard
Calculation amount   ○ Small   ○ Normal   ● Large
1) $\left\{ \sqrt[n]{n!} \right\}$
2) $\left\{ \frac{n}{n^2 + 1} + \frac{n}{n^2 + 2} + \cdots + \frac{n}{n^2 + n} \right\}$
3) $\left\{ \sqrt[n]{3^n + 2^n} \right\}$
4) $\left\{ \frac{a^n - b^n}{a^n + b^n} \right\}, a > 0, \; b > 0$

3.18. Which one of the following series is convergent?

$$S_1 = \sum_{n=1}^{\infty} \left(\frac{n-1}{n}\right)^n$$

$$S_2 = \sum_{n=1}^{\infty} \left(\frac{n+1}{n}\right)^n$$

$$S_3 = \sum_{n=1}^{\infty} \left(\frac{n+1}{n}\right)^{n^2}$$

$$S_4 = \sum_{n=1}^{\infty} \frac{1}{n\sqrt{n}}$$

Difficulty level   ○ Easy   ○ Normal   ● Hard
Calculation amount   ○ Small   ○ Normal   ● Large
1) S_1
2) S_2
3) S_3
4) S_4

References

1. Rahmani-Andebili, M. (2023). Calculus I (2nd Ed.) – Practice Problems, Methods, and Solutions, Springer Nature, 2021.
2. Rahmani-Andebili, M. (2021). Calculus – Practice Problems, Methods, and Solutions, Springer Nature, 2021.
3. Rahmani-Andebili, M. (2021). Precalculus – Practice Problems, Methods, and Solutions, Springer Nature, 2021.

Solutions of Problems: Sequences and Series and Their Applications

4

Abstract

In this chapter, the problems of the third chapter are fully solved, in detail, step-by-step, and with different methods.

4.1. From the concept of growth rate, for $n \longrightarrow \infty$, a, $b > 1$ and $k > 0$, we know that the order of growth rate is as follows [1–3].

$$\log_a n < n^k < b^n < n! < n^n$$

Thus, for this problem, we have:

$$\lim_{n \to \infty} \frac{2^n}{(n+2)!} \sim \lim_{n \to \infty} \frac{1}{(n+2)!} = \frac{1}{\infty!} = 0$$

$$\Rightarrow \lim_{n \to \infty} \frac{2^n}{(n+2)!} = 0$$

Choice (1) is the answer.

4.2. The problem can be solved as follows.

$$\lim_{n \to \infty} \sqrt[8]{\frac{n! + 2n^8 + \ln n}{n! + 5^n + 4n}} \sim \lim_{n \to +\infty} \sqrt[8]{\frac{n!}{n!}} = \lim_{n \to +\infty} 1 = 1$$

$$\Rightarrow \lim_{n \to \infty} \sqrt[8]{\frac{n! + 2n^8 + \ln n}{n! + 5^n + 4n}} = 1$$

Choice (2) is the answer.

In this problem, the rule below was used.

Based on the concept of growth rate, for $n \longrightarrow \infty$, a, $b > 1$ and $k > 0$, the order of growth rate is as follows.

$$\log_a n < n^k < b^n < n! < n^n$$

4.3. From the concept of equivalent functions, we know that for $a > 0$ and an even k:

$$\lim_{n \to \infty} \sqrt[k]{an^k + bn^{k-1} + \dots} = \lim_{n \to \infty} \sqrt[k]{a} \left| n + \frac{b}{ka} \right|$$

And for an odd k, we have:

$$\lim_{n \to \infty} \sqrt[k]{an^k + bn^{k-1} + \dots} = \lim_{n \to \infty} \sqrt[k]{a} \left(n + \frac{b}{ka} \right)$$

Moreover, based on growth rate, we know that:

$$\lim_{n \to \infty} a_1 n^{k_1} + a_2 n^{k_2} \sim a_1 n^{k_1} \qquad \text{if } k_1 > k_2 > 0$$

Based on the information given in the problem, we have:

$$a_n = \frac{\sqrt{n}}{\sqrt{4n^3 + \sin n + 1}} + \frac{\sqrt{n}}{\sqrt{4n^3 + \sin n + 2}} + \dots + \frac{\sqrt{n}}{\sqrt{4n^3 + \sin n + n}}$$

$$\Rightarrow \lim_{n \to \infty} a_n \sim \lim_{n \to \infty} \left(\frac{\sqrt{n}}{\sqrt{4n^3}} + \frac{\sqrt{n}}{\sqrt{4n^3}} + \dots \frac{\sqrt{n}}{\sqrt{4n^3}} \right) = \lim_{n \to \infty} \left(\frac{1}{2n} + \frac{1}{2n} + \dots + \frac{1}{2n} \right)$$

$$\Rightarrow \lim_{n \to \infty} a_n = \lim_{n \to \infty} n \times \left(\frac{1}{2n} \right)$$

$$\Rightarrow \lim_{n \to \infty} a_n = \frac{1}{2}$$

Choice (1) is the answer.

4.4. Based on the problem, we need to determine the coefficient of x^3 in the Maclaurin expansion of the function below based on the following information.

$$f(0) = 1$$

$$f'(x) = 1 + (f(x))^{10}$$

From Maclaurin series or Maclaurin expansion, we know that $\frac{f'''(0)}{3!}$ is the coefficient of x^3 as can be seen in the following.

$$f(x) = f(0) + f'(0)\frac{x}{1!} + f''(0)\frac{x^2}{2!} + +f'''(0)\frac{x^3}{3!} + \dots + f^{(n)}(0)\frac{x^n}{n!} + \dots$$

Therefore:

$$f'(x) = 1 + (f(x))^{10} \Rightarrow f''(x) = 10f'(x) \times (f(x))^9$$

$$\Rightarrow f'''(x) = 10f''(x) \times (f(x))^9 + 90(f'(x))^2 \times (f(x))^8$$

For $x = 0$:

$$f'(0) = 1 + (f(0))^{10} = 1 + 1 = 2$$

$$f''(0) = 10f'(0) \times (f(0))^9 = 10 \times 2 \times 1^9 = 20$$

$$f'''(0) = 10f''(0) \times (f(0))^9 + 90(f'(0))^2 \times (f(0))^8 = 10 \times 20 \times 1^9 + 90(2)^2 \times (1)^8$$

$$\Rightarrow f'''(0) = 380$$

$$\Rightarrow \text{coefficient of } x^3 = \frac{f'''(0)}{3!} = \frac{380}{6} = \frac{190}{3}$$

Choice (2) is the answer.

4.5. First, we need to find a general term for the sequence as follows.

$$1, -\frac{2}{2}, \frac{3}{4}, \frac{-4}{8}, \frac{5}{16}, \cdots = \frac{n(-1)^{n+1}}{2^{n-1}}$$

From the concept of growth rate, for $n \to \infty$, $a, b > 1$ and $k > 0$, we know that the order of growth rate is as follows.

$$\log_a n < n^k < b^n < n! < n^n$$

Hence:

$$\lim_{n \to \infty} \frac{n(-1)^{n+1}}{2^{n-1}} = \lim_{n \to \infty} \frac{1}{2^{n-1}} = \frac{1}{2^{\infty-1}} = \frac{1}{\infty}$$

$$\Rightarrow \lim_{n \to \infty} \frac{n(-1)^{n+1}}{2^{n-1}} = 0$$

Choice (1) is the answer.

4.6. Based on the information given in the problem, we have:

$$S = \sum_{n=1}^{+\infty} \frac{1 + \cos n}{n^2}$$

As we know, the series below, called P-series, is convergent if $P > 1$; otherwise, it is divergent.

$$\sum_{n=1}^{\infty} \frac{1}{n^P}$$

Based on the above-mentioned rule, the series below is convergent.

$$S_0 = \sum_{n=1}^{\infty} \frac{2}{n^2}$$

Since S_0 is convergent and the relation below is held for every term, the series of S is convergent.

$$\frac{1 + \cos n}{n^2} \le \frac{2}{n^2}$$

Choice (1) is the answer.

In this problem, the theorems below were used.

Theorem: The P-series, presented below, is convergent for $P > 1$ and it is divergent for $P \leq 1$.

$$\sum_{n=1}^{\infty} \frac{1}{n^P}$$

Theorem: Suppose that for every term $a_n \leq b_n$. Then, if $\sum_{n=1}^{\infty} a_n$ is divergent, $\sum_{n=1}^{\infty} b_n$ will be divergent as well. Also, if $\sum_{n=1}^{\infty} b_n$ is convergent, $\sum_{n=1}^{\infty} a_n$ will be convergent.

4.7. Based on growth rate, we know that:

$$\lim_{n \to \infty} a_1 n^{k_1} + a_2 n^{k_2} \sim a_1 n^{k_1} \qquad \text{if } k_1 > k_2 > 0$$

Therefore:

$$S = \sum_{n=1}^{+\infty} \frac{10n^2 + 9n + 8}{12n^3 + 11n^2 + 10n + 9} \sim \sum_{n=1}^{+\infty} \frac{10n^2}{12n^3} = \frac{5}{6} \sum_{n=1}^{+\infty} \frac{1}{n}$$

As we know, harmonic series, shown below, is divergent.

$$\sum_{n=1}^{+\infty} \frac{1}{n}$$

Hence, the series of S has a similar behavior, and consequently it is divergent as well.

Choice (2) is the answer.

In this problem, the theorem below was used.

Theorem: Harmonic series is a divergent series presented in the following.

$$\sum_{n=1}^{+\infty} \frac{1}{n}$$

4.8. Based on the problem, we have:

$$S = \sum_{n=1}^{\infty} \frac{\sqrt{n+1} - \sqrt{n}}{\sqrt{n^2 + n}}$$

$$\Rightarrow S = \sum_{n=1}^{\infty} \frac{\sqrt{n+1} - \sqrt{n}}{\sqrt{n}\sqrt{n+1}} = \sum_{n=1}^{\infty} \left(\frac{1}{\sqrt{n}} - \frac{1}{\sqrt{n+1}} \right)$$

As can be noticed, the series is a telescoping series. Therefore:

$$\Rightarrow S = \frac{1}{\sqrt{1}} - \frac{1}{\sqrt{\infty + 1}} = 1 - 0$$

$$\Rightarrow S = 1$$

Choice (2) is the answer.

In this problem, the rule below was used.

Telescoping series, presented below, is a series in which pairs of consecutive terms cancel each other so that only the initial and final terms are left.

$$\sum_{n=1}^{N}(a_n - a_{n+1}) = a_1 - a_{N+1}$$

4.9. Based on the problem, we have:

$$S = \sum_{n=1}^{\infty} n^{1390} x^n$$

To determine the convergence radius (R) and convergence range of a power series in the form of $\sum_{n=1}^{\infty} a_n(x-c)^n$, we can use two methods below.

$$\frac{1}{R} = \lim_{n \to \infty} \sqrt[n]{|a_n|}$$

$$\frac{1}{R} = \lim_{n \to \infty} \left| \frac{a_{n+1}}{a_n} \right|$$

The convergence range can be determined as follows.

$$|x| \le R \;\Rightarrow\; -R \le x \le R$$

For this problem, we can use the second method as follows.

$$\frac{1}{R} = \lim_{n \to \infty} \left| \frac{a_{n+1}}{a_n} \right| = \lim_{n \to \infty} \left| \frac{(n+1)^{1390}}{n^{1390}} \right| \sim \lim_{n \to \infty} \left| \frac{n^{1390}}{n^{1390}} \right| = 1$$

$$\Rightarrow R = 1$$

$$\Rightarrow |x| \le 1 \Rightarrow -1 \le x \le 1$$

It should be noted that for $x = \pm 1$, the series do not have the necessary convergence criterion. Hence:

$$-1 < x < 1$$

Choice (2) is the answer.

4.10. Based on the problem, we have:

$$S = \sum_{n=0}^{\infty} \frac{n!x^n}{3^{n^2}}$$

To determine the convergence radius (R) and convergence range of a power series in the form of $\sum_{n=1}^{\infty} a_n(x-c)^n$, we can use two methods below.

$$\frac{1}{R} = \lim_{n \to \infty} \sqrt[n]{|a_n|}$$

$$\frac{1}{R} = \lim_{n \to \infty} \left| \frac{a_{n+1}}{a_n} \right|$$

The convergence range can be determined as follows.

$$|x| \le R \Rightarrow -R \le x \le R$$

For this problem, we can use the first method as follows.

$$\frac{1}{R} = \lim_{n \to \infty} \sqrt[n]{\left| \frac{n!}{3^{n^2}} \right|} = \lim_{n \to \infty} \frac{\sqrt[n]{n!}}{\sqrt[n]{3^{n^2}}}$$

By using Stirling's approximation, we have:

$$\Rightarrow \frac{1}{R} = \lim_{n \to \infty} \frac{\frac{n}{e}}{3^n}$$

Using the concept of growth rate:

$$\Rightarrow \frac{1}{R} = \frac{1}{e} \lim_{n \to \infty} \frac{n}{3^n} = 0$$

$$\Rightarrow R = \infty$$

Choice (4) is the answer.

In this problem, the rules below were used.

Stirling's approximation:

$$\lim_{n \to +\infty} \sqrt[n]{n!} = \lim_{n \to +\infty} \frac{n}{e}$$

The concept of growth rate states that for $n \to \infty$, $a, b > 1$ and $k > 0$, the order of growth rate is as follows.

$$\log_a n < n^k < b^n < n! < n^n$$

4.11. Based on the problem, we have:

$$S = \sum_{n=1}^{\infty} \frac{(n!)x^n}{(n+1)^n}$$

To determine the convergence radius (R) and convergence range of a power series in the form of $\sum_{n=1}^{\infty} a_n(x-c)^n$, we can use two methods below.

$$\frac{1}{R} = \lim_{n \to \infty} \sqrt[n]{|a_n|}$$

$$\frac{1}{R} = \lim_{n \to \infty} \left| \frac{a_{n+1}}{a_n} \right|$$

The convergence range can be determined as follows.

$$|x| \leq R \Rightarrow -R \leq x \leq R$$

For this problem, we can use the first method as follows.

$$\frac{1}{R} = \lim_{n \to +\infty} \sqrt[n]{\frac{n!}{(n+1)^n}} = \lim_{n \to +\infty} \frac{\sqrt[n]{n!}}{n+1}$$

Based on Stirling's approximation, we have:

$$\Rightarrow \frac{1}{R} = \lim_{n \to +\infty} \frac{\frac{n}{e}}{n+1} = \frac{1}{e} \lim_{n \to +\infty} \frac{n}{n+1}$$

$$\Rightarrow \frac{1}{R} = \frac{1}{e}$$

$$\Rightarrow R = e$$

Choice (2) is the answer.

In this problem, the rule below was used.

Stirling's approximation:

$$\lim_{n \to +\infty} \sqrt[n]{n!} = \lim_{n \to +\infty} \frac{n}{e}$$

4.12. Based on the problem, we need to determine the Taylor series of the function below around $x = 2$.

$$f(x) = \frac{1}{5-x}$$

The Taylor series or Taylor expansion of the function of $f(x)$ around $x = a$ can be calculate as follows.

$$f(x) = f(a) + f'(a)\frac{(x-a)}{1!} + f''(a)\frac{(x-a)^2}{2!} + \ldots + f^{(n)}(a)\frac{(x-a)^n}{n!} + \ldots$$

Moreover, if in the Taylor series $a = 0$, the series is called Maclaurin series or Maclaurin expansion.

$$f(x) = f(0) + f'(0)\frac{x}{1!} + f''(0)\frac{x^2}{2!} + \dots + f^{(n)}(0)\frac{x^n}{n!} + \dots$$

However, the problem can be easily solved without using the direct formula of Taylor series as follows.

$$f(x) = \frac{1}{5-x} = \frac{-1}{x-5} = \frac{-1}{(x-2)-3}$$

$$\Rightarrow f(x) = \frac{-1}{-3\left(1 - \frac{x-2}{3}\right)} = \frac{1}{3}\frac{1}{1 - \left(\frac{x-2}{3}\right)}$$

As we know:

$$\frac{1}{1-x} = 1 + x + x^2 + \dots = \sum_{n=0}^{\infty} x^n$$

Therefore:

$$\Rightarrow f(x) = \frac{1}{3}\sum_{n=0}^{\infty}\left(\frac{x-2}{3}\right)^n = \frac{1}{3} + \frac{(x-2)}{3^2} + \frac{(x-2)^2}{3^3} + \dots$$

Choice (1) is the answer.

4.13. From the concept of equivalent functions, we know that for $a > 0$ and even k:

$$\lim_{n \to \infty} \sqrt[k]{an^k + bn^{k-1} + \dots} = \lim_{n \to \infty} \sqrt[k]{a}\left|n + \frac{b}{ka}\right|$$

And for odd k, we have:

$$\lim_{n \to \infty} \sqrt[k]{an^k + bn^{k-1} + \dots} = \lim_{n \to \infty} \sqrt[k]{a}\left(n + \frac{b}{ka}\right)$$

Moreover, based on growth rate, we know that:

$$\lim_{n \to \infty} a_1 n^{k_1} + a_2 n^{k_2} \sim a_1 n^{k_1} \qquad \text{if } k_1 > k_2 > 0$$

Therefore, for this problem, we have:

$$\lim_{n \to \infty} a_n = \lim_{n \to \infty}\left(\sqrt[3]{n^2 - n^3} + n\right) = \lim_{n \to \infty}\left(-\sqrt[3]{n^3 - n^2} + n\right) \sim \lim_{n \to \infty}\left(-\left(n + \frac{1}{3(-1)}\right) + n\right) = \lim_{n \to \infty}\left(\frac{1}{3}\right)$$

$$\Rightarrow \lim_{n \to \infty} a_n = \frac{1}{3}$$

$$\lim_{n \to \infty} b_n = \lim_{n \to \infty}\frac{\sqrt{n^2 + 1} - \sqrt{n}}{\sqrt[4]{n^3 + n} + \sqrt{n}} \sim \lim_{n \to \infty}\frac{n - \sqrt{n}}{\sqrt[4]{n^3} + \sqrt{n}} \sim \lim_{n \to \infty}\frac{n}{\sqrt[4]{n^3}} = \lim_{n \to \infty}\sqrt[4]{n} = \sqrt[4]{\infty}$$

$$\Rightarrow \lim_{n \to \infty} b_n = \infty$$

Choice (3) is the answer.

4.14. Based on the information given in the problem, we have:

$$a_n = \frac{1 + \frac{1}{2} + \frac{1}{3} + \cdots + \frac{1}{n^3}}{\ln n} \tag{1}$$

$$\lim_{n \to +\infty} a_n = \lim_{n \to +\infty} \frac{1 + \frac{1}{2} + \frac{1}{3} + \cdots + \frac{1}{n^3}}{\ln n} = \lim_{n \to +\infty} \frac{1}{\ln n} \times \sum_{k=1}^{n^3} \frac{1}{k} \tag{2}$$

From the concept of equivalent functions, we know that if $n \to +\infty$:

$$\sum_{k=1}^{n} \frac{1}{k} \sim \ln n \tag{3}$$

Hence:

$$\sum_{k=1}^{n^3} \frac{1}{k} \sim \ln n^3 \tag{4}$$

Solving (2) and (4):

$$\lim_{n \to +\infty} a_n \sim \lim_{n \to \infty} \frac{\ln n^3}{\ln n} = \lim_{n \to \infty} \frac{3 \ln n}{\ln n} = \lim_{n \to \infty} 3$$

$$\Rightarrow \lim_{n \to \infty} a_n = 3$$

Choice (3) is the answer.

In this problem, the rule below was used.

$$\ln a^b = b \ln a$$

4.15. Based on the information given in the problem, we have:

$$a_n = \frac{1}{\sqrt{n^2 + 1}} + \frac{1}{\sqrt{n^2 + 2}} + \cdots \frac{1}{\sqrt{n^2 + n}}$$

Each term of the sequence is equal or larger than $\frac{1}{\sqrt{n^2+n}}$ and equal or smaller than $\frac{1}{\sqrt{n^2+1}}$. In other words:

$$\frac{1}{\sqrt{n^2 + n}} \leq a_n \leq \frac{1}{\sqrt{n^2 + 1}}, \forall n \in \mathbb{N}$$

Therefore, the value of $\lim_{n \to +\infty} a_n$ is larger than $\frac{n}{\sqrt{n^2+n}}$ and smaller than $\frac{n}{\sqrt{n^2+1}}$. In other words:

$$\lim_{n \to +\infty} \frac{n}{\sqrt{n^2 + n}} \leq \lim_{n \to +\infty} a_n \leq \lim_{n \to +\infty} \left(\frac{n}{\sqrt{n^2 + 1}} \right)$$

Based on the concept of equivalent functions, we have:

$$\lim_{n \to +\infty} \frac{n}{\left| n + \frac{1}{2} \right|} \leq \lim_{n \to +\infty} a_n \leq \lim_{n \to +\infty} \left(\frac{n}{|n|} \right)$$

$$\Rightarrow 1 \leq \lim_{n \to +\infty} a_n \leq 1$$

Thus, based on Sandwich theorem, we have:

$$\lim_{n \to +\infty} a_n = 1$$

Choice (3) is the answer.

In this problem, the rules below were used.

From the concept of equivalent functions, we know that for $a > 0$ and an even k:

$$\lim_{n \to \infty} \sqrt[k]{an^k + bn^{k-1} + \dots} = \lim_{n \to \infty} \sqrt[k]{a}\left|n + \frac{b}{ka}\right|$$

And for an odd k, we have:

$$\lim_{n \to \infty} \sqrt[k]{an^k + bn^{k-1} + \dots} = \lim_{n \to \infty} \sqrt[k]{a}\left(n + \frac{b}{ka}\right)$$

Theorem: Sandwich theorem states that if we have:

$$b_n \leq a_n \leq c_n$$

$$\lim_{n \to +\infty} b_n = L$$

$$\lim_{n \to +\infty} c_n = L$$

Then:

$$\lim_{n \to +\infty} a_n = L$$

4.16. Based on the problem, we have:

$$S = \frac{1}{1} + \frac{1}{1+2} + \frac{1}{1+2+3} + \frac{1}{1+2+3+4} + \cdots$$

$$\Rightarrow S = \sum_{n=1}^{\infty} \frac{1}{\frac{n(n+1)}{2}} = \sum_{n=1}^{\infty} \frac{2}{n(n+1)}$$

$$\Rightarrow S = 2\sum_{n=1}^{\infty} \left(\frac{1}{n} - \frac{1}{n+1}\right)$$

As can be noticed, the series is a telescoping series. Therefore:

$$S = 2\left(1 - \frac{1}{\infty + 1}\right) = 2(1 - 0)$$

$$\Rightarrow S = 2$$

Choice (1) is the answer.

In this problem, the rules below were used.

$$1 + 2 + 3 + \ldots + n = \frac{n(n+1)}{2}$$

Telescoping series, presented below, is a series in which pairs of consecutive terms cancel each other so that only the initial and final terms are left.

$$\sum_{n=1}^{N} (a_n - a_{n+1}) = a_1 - a_{N+1}$$

4.17. A sequence in the form of $a_n = \{a_1, a_2, \ldots\}$ is convergent if its limit when $n \to \infty$ is a unique finite number ($\lim\limits_{n \to +\infty} a_n$); otherwise, the sequence is divergent.

Choice (1): Based on Stirling's approximation, we have:

$$\lim_{n \to +\infty} \sqrt[n]{n!} = \lim_{n \to +\infty} \frac{n}{e}$$

$$\Rightarrow \lim_{n \to +\infty} \sqrt[n]{n!} = \infty$$

Hence, the sequence is divergent.

Choice (2): Each term of the sequence is equal or larger than $\frac{n}{n^2+n}$ and equal or smaller than $\frac{n}{n^2+1}$. In other words:

$$\frac{n}{n^2+n} \le \frac{n}{n^2+1} + \frac{n}{n^2+2} + \cdots + \frac{n}{n^2+n} \le \frac{n}{n^2+1}, \forall n \epsilon \mathbb{N}$$

Therefore, the value of $\lim\limits_{n \to +\infty} a_n$ is larger than $\frac{n^2}{n^2+n}$ and smaller than $\frac{n^2}{n^2+1}$. In other words:

$$\lim_{n \to +\infty} \frac{n^2}{n^2+n} \le \lim_{n \to +\infty} a_n \le \lim_{n \to +\infty} \left(\frac{n^2}{n^2+1} \right)$$

$$\Rightarrow 1 \le \lim_{n \to +\infty} a_n \le 1$$

Therefore, based on Sandwich theorem:

$$\lim_{n \to +\infty} a_n = 1$$

Hence, the sequence is convergent.

Choice (3): Based on growth rate, we know that:

$$\lim_{n \to \infty} a_1{}^n + a_2{}^n \sim a_1{}^n \quad \text{if } a_1 > a_2 > 0$$

Thus:

$$\lim_{n \to +\infty} \sqrt[n]{3^n + 2^n} \sim \lim_{n \to +\infty} \sqrt[n]{3^n} = \lim_{n \to +\infty} 3 = 3$$

$$\Rightarrow \lim_{n \to +\infty} a_n = 3$$

Thus, the sequence is convergent.

Choice (4): Based on growth rate, we know that:

$$\lim_{n \to \infty} a_1{}^n + a_2{}^n \sim a_1{}^n \qquad \text{if } a_1 > a_2 > 0$$

Now, if $a > b$:

$$\lim_{n \to +\infty} \frac{a^n - b^n}{a^n + b^n} \sim \lim_{n \to +\infty} \frac{a^n}{a^n} = 1$$

$$\Rightarrow \lim_{n \to +\infty} a_n = 1$$

If $b > a$:

$$\lim_{n \to +\infty} \frac{a^n - b^n}{a^n + b^n} \sim \lim_{n \to +\infty} \frac{-b^n}{b^n} = -1$$

$$\Rightarrow \lim_{n \to +\infty} a_n = -1$$

Therefore, in any condition, the sequence is convergent.

Choice (1) is the answer.

In this problem, the rule below was used.

Theorem: Sandwich theorem states that if we have:

$$b_n \leq a_n \leq c_n$$

$$\lim_{n \to +\infty} b_n = L$$

$$\lim_{n \to +\infty} c_n = L$$

Then:

$$\lim_{n \to +\infty} a_n = L$$

4.18. The problem can be solved by using the following theorem.

Theorem: It states that the necessary, but not enough, criterion for a series in the form of $S_n = \sum_{n=1}^{+\infty} a_n$ to be convergent is that the limit of its general term must be zero when $n \to \infty$. In other words:

$$\lim_{n \to +\infty} a_n = 0$$

Choice (1):

$$\lim_{n \to \infty} \left(\frac{n-1}{n}\right)^n = \lim_{n \to \infty} \left(1 - \frac{1}{n}\right)^n = e^{-1} \neq 0$$

Choice (2):

$$\lim_{n \to \infty} \left(\frac{n+1}{n}\right)^n = \lim_{n \to \infty} \left(1 + \frac{1}{n}\right)^n = e \neq 0$$

Choice (3):

$$\lim_{n \to \infty} \left(\frac{n+1}{n}\right)^{n^2} = \lim_{n \to \infty} \left(\left(\frac{n+1}{n}\right)^n\right)^n = \lim_{n \to \infty} \left(\left(1 + \frac{1}{n}\right)^n\right)^n = e^\infty = \infty \neq 0$$

Choice (4):

$$\lim_{n \to \infty} \frac{1}{n\sqrt{n}} = \lim_{n \to \infty} \frac{1}{n^{\frac{3}{2}}} = \frac{1}{\infty} = 0$$

Thus, the series of S_4 has the necessary criterion of convergence.

On the other hand, we know that P-series, shown below, is convergent for $P > 1$.

$$\sum_{n=1}^{\infty} \frac{1}{n^P}$$

For Choice (4), $P = 1.5 > 1$; therefore, the series is convergent. Choice (4) is the answer.

In this problem, the rule below was used.

$$\lim_{n \to \infty} \left(1 + \frac{a}{bn}\right)^{cn} = e^{\frac{ac}{b}}$$

References

1. Rahmani-Andebili, M. (2023). Calculus I (2nd Ed.) – Practice Problems, Methods, and Solutions, Springer Nature, 2021.
2. Rahmani-Andebili, M. (2021). Calculus – Practice Problems, Methods, and Solutions, Springer Nature, 2021.
3. Rahmani-Andebili, M. (2021). Precalculus – Practice Problems, Methods, and Solutions, Springer Nature, 2021.

Problems: Polar Coordinate System

Abstract

In this chapter, the basic and advanced problems concerned with polar coordinate system are presented. The subjects include tangent line on a curve, radius of curve, polar equation, spiral, transferring from cartesian coordinate to polar coordinate and vice versa, and curve types such as ellipse, straight line, parabola, hyperbola. In this chapter, the problems are categorized in different levels based on their difficulty levels (easy, normal, and hard) and calculation amounts (small, normal, and large). Additionally, the problems are ordered from the easiest problem with the smallest computations to the most difficult problems with the largest calculations.

5.1. Express the cartesian position of the point of $\left(\frac{1}{2}, -\frac{\sqrt{3}}{2}\right)$ in polar coordinate system (r, θ) [1–3].

Difficulty level ● Easy ○ Normal ○ Hard
Calculation amount ● Small ○ Normal ○ Large

1) $\left(2, \frac{\pi}{3}\right)$

2) $\left(1, \frac{\pi}{6}\right)$

3) $\left(1, -\frac{\pi}{6}\right)$

4) $\left(1, -\frac{\pi}{3}\right)$

5.2. Express the cartesian position of the point of $\left(-3, \sqrt{3}\right)$ in polar coordinate system (r, θ).

Difficulty level ● Easy ○ Normal ○ Hard
Calculation amount ● Small ○ Normal ○ Large

1) $\left(2\sqrt{3}, \frac{5\pi}{6}\right)$

2) $\left(3\sqrt{3}, \frac{5\pi}{6}\right)$

3) $\left(2\sqrt{3}, \frac{\pi}{6}\right)$

4) $\left(2\sqrt{3}, \frac{\pi}{3}\right)$

5.3. Express the polar position of the point of $r = 1, \theta = 0$ in cartesian coordinate system (x, y).

Difficulty level ● Easy ○ Normal ○ Hard
Calculation amount ● Small ○ Normal ○ Large

1) $(1, 1)$

2) $(1, 0)$

3) (0, 1)

4) (0, 0)

5.4. Express the polar position of the point of $r = \sqrt{2}, \theta = -\dfrac{\pi}{4}$ in cartesian coordinate system (x, y).

Difficulty level   ● Easy   ○ Normal   ○ Hard

Calculation amount   ● Small   ○ Normal   ○ Large

1) $\left(\sqrt{2}, \sqrt{2}\right)$

2) $\left(\dfrac{\sqrt{2}}{2}, \dfrac{\sqrt{2}}{2}\right)$

3) $(1, -1)$

4) $(1, 1)$

5.5. Calculate the amount of angle between the tangent line and the radius of the curve of $r = \theta^2 + 1$ at $\theta = 1$.

Difficulty level   ○ Easy   ○ Normal   ● Hard

Calculation amount   ● Small   ○ Normal   ○ Large

1) $\dfrac{\pi}{6}$

2) $\dfrac{\pi}{4}$

3) $\dfrac{\pi}{3}$

4) $\dfrac{\pi}{2}$

5.6. Calculate the amount of angle between the tangent line and the radius of the curve of $r = 2 + 2\sin\theta$ at the point of $(3, \dfrac{\pi}{6})$.

Difficulty level   ○ Easy   ○ Normal   ● Hard

Calculation amount   ● Small   ○ Normal   ○ Large

1) $\dfrac{\pi}{6}$

2) $\dfrac{\pi}{4}$

3) $\dfrac{\pi}{3}$

4) $\dfrac{\pi}{2}$

5.7. Determine the polar equation of a circle with the center on positive side of x-axis and radius of two while passing from the origin.

Difficulty level   ○ Easy   ○ Normal   ● Hard

Calculation amount   ● Small   ○ Normal   ○ Large

1) $r = 4\cos\theta$

2) $r = 2\cos\theta$

3) $r = 4\sin\theta$

4) $r = 2\sin\theta$

5.8. Calculate the surface area of the spiral of $r = e^{\frac{\theta}{4\pi}}$ from $\theta = 0$ to $\theta = 2\pi$.

Difficulty level   ○ Easy   ○ Normal   ● Hard

Calculation amount   ● Small   ○ Normal   ○ Large

1) $\pi(e - 1)$

2) $2\pi\left(\sqrt{e} - 1\right)$

3) $2\pi(e - 1)$

4) $4\pi\left(\sqrt{e} - 1\right)$

5.9. Calculate the surface area that the curve of $r = (4 - 4\sin\theta\cos\theta)^{\frac{1}{2}}$ separates from the first quadrant.

Difficulty level   ○ Easy   ○ Normal   ● Hard

Calculation amount   ● Small   ○ Normal   ○ Large

1) $\pi - 1$
2) π
3) $\pi + 1$
4) $\pi - 2$

5.10. What is the curve type of the polar function of $r = \cot\theta\csc\theta$.

Difficulty level   ○ Easy   ○ Normal   ● Hard

Calculation amount   ○ Small   ● Normal   ○ Large

1) Ellipse
2) Straight line
3) Parabola
4) Hyperbola

5.11. Determine the curve type of the polar function below.

$$r = \frac{5}{3\cos\theta + 2\sin\theta}$$

Difficulty level   ○ Easy   ○ Normal   ● Hard

Calculation amount   ○ Small   ● Normal   ○ Large

1) Straight line
2) Ellipse
3) Parabola
4) Hyperbola

5.12. Express the function of $\cos 2\theta = 1$ in cartesian coordinate system.

Difficulty level   ○ Easy   ○ Normal   ● Hard

Calculation amount   ○ Small   ● Normal   ○ Large

1) $x - y = 0$
2) $x + y = 0$
3) $x = 0$
4) $y = 0$

5.13. Express the function below in cartesian coordinate system.

$$\begin{cases} x = \cos^2\theta \\ y = \sin\theta\cos\theta \end{cases}$$

Difficulty level   ○ Easy   ○ Normal   ● Hard

Calculation amount   ○ Small   ● Normal   ○ Large

1) $\left(x - \frac{1}{2}\right)^2 + y^2 = \frac{1}{4}$

2) $x^2 + \left(y - \frac{1}{2}\right)^2 = \frac{1}{4}$

3) $\left(x + \frac{1}{2}\right)^2 + y^2 = \frac{1}{4}$

4) $\left(x - \frac{1}{2}\right)^2 + \left(y - \frac{1}{2}\right)^2 = \frac{1}{4}$

5.14. Calculate the amount of angle between the curves of $r = 2(1 + \sin\theta)$ and $r = 3(1 - \sin\theta)$ at the intersection point.

Difficulty level   ○ Easy   ○ Normal   ● Hard

Calculation amount   ○ Small   ● Normal   ○ Large

1) $\dfrac{\pi}{2}$

2) $\dfrac{\pi}{4}$

3) $\arctan\dfrac{2}{3}$

4) $\arctan\dfrac{1}{7}$

5.15. Determine the curve type of the polar function below.

$$r = \frac{4}{2 - \cos\theta}$$

Difficulty level   ○ Easy   ○ Normal   ● Hard

Calculation amount   ○ Small   ● Normal   ○ Large

1) Ellipse
2) Parabola
3) Hyperbola
4) Two crossing straight lines

5.16. Determine the number of intersection points of the following polar functions.

$$r = \frac{\cos\theta}{\sin^2\theta}$$

$$r = -\frac{3}{\cos\theta + 4\sin\theta}$$

Difficulty level   ○ Easy   ○ Normal   ● Hard

Calculation amount   ○ Small   ○ Normal   ● Large

1) 1
2) 2
3) ∞
4) 0

References

1. Rahmani-Andebili, M. (2023). Calculus I (2nd Ed.) – Practice Problems, Methods, and Solutions, Springer Nature, 2021.
2. Rahmani-Andebili, M. (2021). Calculus – Practice Problems, Methods, and Solutions, Springer Nature, 2021.
3. Rahmani-Andebili, M. (2021). Precalculus – Practice Problems, Methods, and Solutions, Springer Nature, 2021.

Abstract

In this chapter, the problems of the fifth chapter are fully solved, in detail, step-by-step, and with different methods.

6.1. Based on the information given in the problem, we have [1–3]:

$$(x, y) = \left(\frac{1}{2}, -\frac{\sqrt{3}}{2} \right)$$

The relations below are used to transfer from cartesian coordinate to polar coordinate.

$$r = \sqrt{x^2 + y^2}$$

$$\theta = \tan^{-1}\left|\frac{y}{x}\right| \quad \text{if } x > 0, y > 0$$

$$\theta = \pi - \tan^{-1}\left|\frac{y}{x}\right| \quad \text{if } x < 0, y > 0$$

$$\theta = \pi + \tan^{-1}\left|\frac{y}{x}\right| \quad \text{if } x < 0, y < 0$$

$$\theta = -\tan^{-1}\left|\frac{y}{x}\right| \quad \text{if } x > 0, y < 0$$

Therefore, for this problem, we have:

$$r = \sqrt{\left(\frac{1}{2}\right)^2 + \left(-\frac{\sqrt{3}}{2}\right)^2} = \sqrt{\frac{1}{4} + \frac{3}{4}} = 1$$

$$\theta = -\tan^{-1}\frac{\frac{\sqrt{3}}{2}}{\frac{1}{2}} = -\tan^{-1}\sqrt{3} = -\frac{\pi}{3}$$

$$\Rightarrow (r, \theta) = \left(1, -\frac{\pi}{3} \right)$$

Choice (4) is the answer.

In this problem, the rule below was used.

$$\tan^{-1}\sqrt{3} = \frac{\pi}{3}$$

6.2. Based on the information given in the problem, we have:

$$(x, y) = \left(-3, \sqrt{3}\right)$$

The relations below are used to transfer from cartesian coordinate to polar coordinate.

$$r = \sqrt{x^2 + y^2}$$

$$\theta = \tan^{-1}\left|\frac{y}{x}\right| \quad \text{if } x > 0, y > 0$$

$$\theta = \pi - \tan^{-1}\left|\frac{y}{x}\right| \quad \text{if } x < 0, y > 0$$

$$\theta = \pi + \tan^{-1}\left|\frac{y}{x}\right| \quad \text{if } x < 0, y < 0$$

$$\theta = -\tan^{-1}\left|\frac{y}{x}\right| \quad \text{if } x > 0, y < 0$$

Therefore, for this problem, we have:

$$r = \sqrt{(-3)^2 + \left(\sqrt{3}\right)^2} = \sqrt{9 + 3} = 2\sqrt{3}$$

$$\theta = \pi - \tan^{-1}\frac{\sqrt{3}}{3} = \pi - \frac{\pi}{6} = \frac{5\pi}{6}$$

$$\Rightarrow (r, \theta) = \left(2\sqrt{3}, \frac{5\pi}{6}\right)$$

Choice (1) is the answer.

In this problem, the rule below was used.

$$\tan^{-1}\frac{\sqrt{3}}{3} = \frac{\pi}{6}$$

6.3. Based on the information given in the problem, we have:

$$(r, \theta) = (1, 0)$$

The relations below are used to transfer from polar coordinate to cartesian coordinate.

$$x = r\cos\theta$$

$$y = r\sin\theta$$

Therefore, for this problem, we have:

$$x = 1 \cos 0 = 1$$

$$y = 1 \sin 0 = 0$$

$$\Rightarrow (x, y) = (1, 0)$$

Choice (2) is the answer.

In this problem, the rules below were used.

$$\cos 0 = 1$$

$$\sin 0 = 0$$

6.4. Based on the information given in the problem, we have:

$$(r, \theta) = \left(\sqrt{2}, -\frac{\pi}{4} \right)$$

The relations below are used to transfer from polar coordinate to cartesian coordinate.

$$x = r \cos \theta$$

$$y = r \sin \theta$$

Therefore, for this problem, we have:

$$x = \sqrt{2} \cos \left(-\frac{\pi}{4} \right) = 1$$

$$y = \sqrt{2} \sin \left(-\frac{\pi}{4} \right) = -1$$

$$\Rightarrow (x, y) = (1, -1)$$

Choice (3) is the answer.

In this problem, the rules below were used.

$$\cos \left(-\frac{\pi}{4} \right) = \frac{\sqrt{2}}{2}$$

$$\sin \left(-\frac{\pi}{4} \right) = -\frac{\sqrt{2}}{2}$$

6.5. The amount of angle between the tangent line on a polar curve and the radius of the curve can be calculated as follows.

$$\tan \alpha = \frac{f(\theta)}{f'(\theta)} = \frac{r(\theta)}{\frac{dr(\theta)}{d\theta}} \tag{1}$$

Based on the information given in the problem, we have:

$$r = \theta^2 + 1 \tag{2}$$

$$\theta = 1 \tag{3}$$

Solving (1)–(3):

$$\tan \alpha = \left. \frac{\theta^2 + 1}{2\theta} \right|_{\theta = 1} = 1$$

$$\Rightarrow \alpha = \tan^{-1}(1) = \frac{\pi}{4}$$

Choice (2) is the answer.

In this problem, the rule below was used.

$$\tan^{-1}(1) = \frac{\pi}{4}$$

6.6. Based on the information given in the problem, we have:

$$r = 2 + 2\sin\theta \tag{1}$$

$$(r, \theta) = \left(3, \frac{\pi}{6}\right) \tag{2}$$

The amount of angle between the tangent line on a polar curve and the radius of the curve can be calculated as follows.

$$\tan \alpha = \frac{f(\theta)}{f'(\theta)} = \frac{r(\theta)}{\frac{dr(\theta)}{d\theta}} \tag{3}$$

Solving (1)–(3):

$$\tan \alpha = \left. \frac{2 + 2\sin\theta}{2\cos\theta} \right|_{(r,\theta) = \left(3, \frac{\pi}{6}\right)} = \frac{2 + 2\sin\frac{\pi}{6}}{2\cos\frac{\pi}{6}} = \frac{3}{\sqrt{3}} = \sqrt{3}$$

$$\Rightarrow \alpha = \tan^{-1}\left(\sqrt{3}\right) = \frac{\pi}{3}$$

Choice (3) is the answer.

In this problem, the rules below were used.

$$\frac{d}{dx}\sin x = \cos x$$

$$\cos\frac{\pi}{6} = \frac{\sqrt{3}}{2}$$

$$\sin\frac{\pi}{6} = \frac{1}{2}$$

$$\tan^{-1}\left(\sqrt{3}\right) = \frac{\pi}{3}$$

6.7. The polar equations of a circle with the center on positive and negative sides of x-axis and radius of "*a*" while passing from the origin are as follows, respectively.

$$r = 2a \cos \theta$$

$$r = -2a \cos \theta$$

Moreover, the polar equations of a circle with the center on positive and negative sides of y-axis and radius of "*a*" while passing from the origin are as follows, respectively.

$$r = 2a \sin \theta$$

$$r = -2a \sin \theta$$

Thus, for this problem, we have:

$$r = 4 \cos \theta$$

Choice (1) is the answer.

6.8. The surface area of a spiral in the form of $r = f(\theta)$ from θ_1 to θ_2 can be calculated as follows.

$$S = \frac{1}{2} \int_{\theta_1}^{\theta_2} r^2 d\theta$$

Based on the information given in the problem, we have:

$$r = e^{\frac{\theta}{4\pi}} \tag{1}$$

$$0 \le \theta \le 2\pi \tag{2}$$

For this problem, we have:

$$S = \frac{1}{2} \int_0^{2\pi} e^{\frac{\theta}{2\pi}} d\theta$$

$$\Rightarrow S = \pi e^{\frac{\theta}{2\pi}} \bigg|_0^{2\pi} = \pi \left(e^1 - e^0 \right)$$

$$\Rightarrow S = \pi (e - 1)$$

Choice (1) is the answer.

In this problem, the rule below was used.

$$\int e^{ax} dx = \frac{1}{a} e^{ax} + c$$

6.9. The surface area of a spiral in the form of $r = f(\theta)$ from θ_1 to θ_2 can be calculated as follows.

$$S = \frac{1}{2}\int_{\theta_1}^{\theta_2} r^2 d\theta$$

Based on the information given in the problem, we have:

$$r = (4 - 4\sin\theta\cos\theta)^{\frac{1}{2}} \tag{1}$$

$$0 \le \theta \le \frac{\pi}{2} \tag{2}$$

Thus, for this problem, we have:

$$\Rightarrow S = \frac{1}{2}\int_{\theta_1}^{\theta_2} r^2 d\theta = \frac{1}{2}\int_0^{\frac{\pi}{2}}(4 - 4\sin\theta\cos\theta)d\theta$$

$$\Rightarrow S = \frac{1}{2}\left(4\theta - 2\sin^2\theta\right)\Big|_0^{\frac{\pi}{2}} = \frac{1}{2}\left[\left(4 \times \frac{\pi}{2} - 2\sin^2\frac{\pi}{2}\right) - (0 - 0)\right]$$

$$\Rightarrow S = \pi - 1$$

Choice (1) is the answer.

In this problem, the rule below was used.

$$\int u^n du = \frac{1}{n+1}u^{n+1} + c$$

6.10. Based on the information given in the problem, we have:

$$r = \cot\theta \csc\theta \tag{1}$$

It can be simplified as follows.

$$r = \left(\frac{\cos\theta}{\sin\theta}\right)\left(\frac{1}{\sin\theta}\right) = \frac{\cos\theta}{\sin^2\theta} \tag{2}$$

By transferring from polar coordinate to cartesian coordinate, we have:

$$y = r\sin\theta \tag{3}$$

$$x = r\cos\theta \tag{4}$$

$$x^2 + y^2 = r^2 \tag{5}$$

Solving (2)–(5):

$$\Rightarrow \sqrt{x^2 + y^2} = \frac{\dfrac{x}{\sqrt{x^2 + y^2}}}{\left(\dfrac{y}{\sqrt{x^2 + y^2}}\right)^2} \Rightarrow \frac{y^2}{\sqrt{x^2 + y^2}} = \frac{x}{\sqrt{x^2 + y^2}}$$

$$\Rightarrow y^2 = x$$

It is the equation of a parabola. Choice (3) is the answer.

In general, the equation of a parabola is as follows.

$$(x - x_0)^2 = 4a(y - y_0)$$

$$(y - y_0)^2 = 4a(x - x_0)$$

In this problem, the rules below were used.

$$\cot\theta = \frac{\cos\theta}{\sin\theta}$$

$$\csc\theta = \frac{1}{\sin\theta}$$

6.11. Based on the information given in the problem, we have:

$$r = \frac{5}{3\cos\theta + 2\sin\theta} \tag{1}$$

$$\Rightarrow 3r\cos\theta + 2r\sin\theta = 5 \tag{2}$$

By transferring from polar coordinate to cartesian coordinate, we have:

$$y = r\sin\theta \tag{3}$$

$$x = r\cos\theta \tag{4}$$

$$x^2 + y^2 = r^2 \tag{5}$$

Solving (2)–(5):

$$3x + 2y = 5$$

It is the equation of a straight line. Choice (1) is the answer.

In general, the equation of a straight line is as follows.

$$ax + by = c$$

6.12. Based on the information given in the problem, we have:

$$\cos 2\theta = 1 \tag{1}$$

$$\Rightarrow \cos^2\theta - \sin^2\theta = 1 \tag{2}$$

By transferring from polar coordinate to cartesian coordinate, we have:

$$y = r \sin \theta \tag{3}$$

$$x = r \cos \theta \tag{4}$$

$$x^2 + y^2 = r^2 \tag{5}$$

Solving (2)–(5):

$$\left(\frac{x}{\sqrt{x^2 + y^2}}\right)^2 - \left(\frac{y}{\sqrt{x^2 + y^2}}\right)^2 = 1 \Rightarrow x^2 - y^2 = x^2 + y^2 \Rightarrow 2y^2 = 0$$

$$\Rightarrow y = 0$$

Choice (4) is the answer.

In this problem, the rule below was used.

$$\cos 2\theta = \cos^2 \theta - \sin^2 \theta$$

6.13. Based on the information given in the problem, we have:

$$\begin{cases} x = \cos^2 \theta \\ y = \sin \theta \cos \theta \end{cases} \tag{1}$$

By transferring from polar coordinate to cartesian coordinate, we have:

$$y = r \sin \theta \tag{2}$$

$$x = r \cos \theta \tag{3}$$

$$x^2 + y^2 = r^2 \tag{4}$$

Solving (1)–(4):

$$\begin{cases} x = \left(\dfrac{x}{\sqrt{x^2 + y^2}}\right)^2 \\ y = \left(\dfrac{y}{\sqrt{x^2 + y^2}}\right)\left(\dfrac{x}{\sqrt{x^2 + y^2}}\right) \end{cases} \Rightarrow \begin{cases} x = \dfrac{x^2}{x^2 + y^2} & (5) \\ y = \dfrac{xy}{x^2 + y^2} & (6) \end{cases}$$

From (5) or (6), we have:

$$x^2 + y^2 = x \Rightarrow x^2 - x + y^2 = 0$$

$$\Rightarrow \left(x - \frac{1}{2}\right)^2 - \frac{1}{4} + y^2 = 0 \Rightarrow \left(x - \frac{1}{2}\right)^2 + y^2 = \frac{1}{4}$$

Choice (1) is the answer.

6.14. Suppose α_1 is the angle between the tangent line and the radius of the curve of $r = f_1(\theta)$ at θ_0 and α_2 is the angle between the tangent line and the radius of the curve of $r = f_2(\theta)$ at θ_0. The acute or straight angle between the two polar curves at the intersection point ($\theta = \theta_0$) can be calculated as follows.

$$\tan \psi = \frac{\tan \alpha_1 - \tan \alpha_2}{1 + \tan \alpha_1 \tan \alpha_2} \tag{1}$$

Moreover, the amount of angle between the tangent line on a polar curve and the radius of the curve can be calculated as follows.

$$\tan \alpha = \frac{f(\theta)}{f'(\theta)} = \frac{r(\theta)}{\frac{dr(\theta)}{d\theta}} \tag{2}$$

Based on the information given in the problem, we have:

$$r = 2(1 + \sin \theta) \tag{1}$$

$$r = 3(1 - \sin \theta) \tag{2}$$

First, we need to find the intersection point of the polar curves as follows.

$$2(1 + \sin \theta) = 3(1 - \sin \theta) \Rightarrow 5 \sin \theta = 1 \Rightarrow \sin \theta = \frac{1}{5} \Rightarrow \cos \theta = \frac{\sqrt{24}}{5}$$

Then:

$$\tan \alpha_1 = \frac{f_2(\theta)}{f'_2(\theta)} = \frac{2(1 + \sin \theta)}{2 \cos \theta} = \frac{1 + \sin \theta}{\cos \theta} = \frac{1 + \frac{1}{5}}{\frac{\sqrt{24}}{5}} = \frac{6}{\sqrt{24}} = \frac{\sqrt{6}}{2}$$

$$\tan \alpha_2 = \frac{f_2(\theta)}{f'_2(\theta)} = \frac{3(1 - \sin \theta)}{- 3 \cos \theta} = \frac{1 - \sin \theta}{- \cos \theta} = \frac{1 - \frac{1}{5}}{- \frac{\sqrt{24}}{5}} = \frac{4}{- \sqrt{24}} = \frac{- \sqrt{6}}{3}$$

Therefore:

$$\Rightarrow \tan \psi = \frac{\frac{\sqrt{6}}{2} - \left(\frac{\sqrt{6}}{3}\right)}{1 + \left(\frac{\sqrt{6}}{2}\right)\left(\frac{-\sqrt{6}}{3}\right)} = \frac{\frac{\sqrt{6}}{6}}{1 - 1} = \infty$$

$$\Rightarrow \psi = \tan^{-1}(\infty) = \frac{\pi}{2}$$

Choice (1) is the answer.

In this problem, the rules below were used.

$$\sin^2 x + \cos^2 x = 1$$

$$\frac{d}{dx} \sin x = \cos x$$

$$\tan^{-1}(\infty) = \frac{\pi}{2}$$

6.15. Based on the information given in the problem, we have:

$$r = \frac{4}{2 - \cos\theta} \tag{1}$$

$$\Rightarrow 2r - r\cos\theta = 4 \tag{2}$$

By transferring from polar coordinate to cartesian coordinate, we have:

$$y = r\sin\theta \tag{3}$$

$$x = r\cos\theta \tag{4}$$

$$x^2 + y^2 = r^2 \tag{5}$$

Solving (2)–(5):

$$2\sqrt{x^2 + y^2} - x = 4 \Rightarrow 2\sqrt{x^2 + y^2} = x + 4$$

$$\Rightarrow 4\left(x^2 + y^2\right) = x^2 + 8x + 16 \Rightarrow 3x^2 - 8x + 4y^2 = 16$$

$$\Rightarrow 3\left(x^2 - \frac{8}{3}x + \frac{16}{9}\right) + 4y^2 = 16 + \frac{16}{3} \Rightarrow 3\left(x - \frac{4}{3}\right)^2 + 4y^2 = \frac{64}{3}$$

$$\Rightarrow \frac{\left(x - \frac{4}{3}\right)^2}{\left(\frac{8}{3}\right)^2} + \frac{(y - 0)^2}{\left(\frac{8}{2\sqrt{3}}\right)^2} = 1$$

It is the equation of an ellipse. Choice (1) is the answer.

In general, the equation of an ellipse is as follows.

$$\frac{(x - x_0)^2}{a^2} + \frac{(y - y_0)^2}{b^2} = 1$$

6.16. Based on the information given in the problem, we have:

$$r = \frac{\cos\theta}{\sin^2\theta} \tag{1}$$

$$r = -\frac{3}{\cos\theta + 4\sin\theta} \tag{2}$$

The problem can be easily solved by transferring from polar coordinate to cartesian coordinate as follows:

$$y = r\sin\theta \tag{3}$$

$$x = r\cos\theta \tag{4}$$

$$x^2 + y^2 = r^2 \tag{5}$$

Solving (1) and (3)–(5):

$$\sqrt{x^2+y^2} = \frac{\frac{x}{\sqrt{x^2+y^2}}}{\left(\frac{y}{\sqrt{x^2+y^2}}\right)^2} \Rightarrow \frac{y^2}{\sqrt{x^2+y^2}} = \frac{x}{\sqrt{x^2+y^2}} \Rightarrow y^2 = x \tag{6}$$

Solving (2) and (3)–(5):

$$\sqrt{x^2+y^2} = -\frac{3}{\frac{x}{\sqrt{x^2+y^2}} + 4\frac{y}{\sqrt{x^2+y^2}}} \Rightarrow x + 4y = -3 \tag{7}$$

Solving (6) and (7):

$$y^2 + 4y = -3 \Rightarrow y^2 + 4y + 3 = 0$$

$$\Rightarrow y = \frac{-4 \pm \sqrt{4^2 - 4 \times 1 \times 3}}{2} = \frac{-4 \pm 2}{2}$$

$$\Rightarrow y = -1, -3$$

Choice (2) is the answer.

In this problem, the rule below was used.

$$ax^2 + bx + c = 0$$

$$\Rightarrow x = \frac{-b \pm \sqrt{b^2 - 4ac}}{2a}$$

References

1. Rahmani-Andebili, M. (2023). Calculus I (2nd Ed.) – Practice Problems, Methods, and Solutions, Springer Nature, 2021.
2. Rahmani-Andebili, M. (2021). Calculus – Practice Problems, Methods, and Solutions, Springer Nature, 2021.
3. Rahmani-Andebili, M. (2021). Precalculus – Practice Problems, Methods, and Solutions, Springer Nature, 2021.

Problems: Complex numbers

Abstract

In this chapter, the basic and advanced problems of complex numbers are presented. The subjects include the operations on complex numbers as well as on functions in complex form. In this chapter, the problems are categorized in different levels based on their difficulty levels (easy, normal, and hard) and calculation amounts (small, normal, and large). Additionally, the problems are ordered from the easiest problem with the smallest computations to the most difficult problems with the largest calculations.

7.1. Calculate the value of e^{0i}.

Difficulty level ● Easy ○ Normal ○ Hard
Calculation amount ● Small ○ Normal ○ Large
1) 0
2) 1
3) −1
4) ∞

7.2. Calculate the value of $e^{\pi i}$ [1–3].

Difficulty level ● Easy ○ Normal ○ Hard
Calculation amount ● Small ○ Normal ○ Large
1) 0
2) 1
3) −1
4) ∞

7.3. Calculate the complex conjugate of i.

Difficulty level ● Easy ○ Normal ○ Hard
Calculation amount ● Small ○ Normal ○ Large
1) $1 - i$
2) 0
3) $1 + i$
4) $-i$

7.4. Which one of the relations below is wrong?

Difficulty level ● Easy ○ Normal ○ Hard
Calculation amount ● Small ○ Normal ○ Large
1) $\overline{(a - ib)} = a + ib$
2) $\overline{\left(\dfrac{z_1}{z_2}\right)} = \dfrac{\overline{z_1}}{\overline{z_2}}$

3) $(a + ib)(a - ib) = a^2 - b^2$

4) $i = \sqrt{-1}$

7.5. Which one of the relations below is wrong?

Difficulty level   ● Easy   ○ Normal   ○ Hard
Calculation amount   ● Small   ○ Normal   ○ Large

1) $|a + ib| = \sqrt{a^2 + b^2}$

2) $Re(z_1 z_2) = x_1 x_2 - y_1 y_2$

3) $Im(z_1 z_2) = x_1 x_2 + y_1 y_2$

4) $e^{ix} = \cos x + i \sin x$

7.6. Which one of the relations below is wrong?

Difficulty level   ● Easy   ○ Normal   ○ Hard
Calculation amount   ● Small   ○ Normal   ○ Large

1) $\dfrac{z_1}{z_2} = \dfrac{|z_1|}{|z_2|} e^{i(\theta_1 - \theta_2)}$

2) $\dfrac{1}{z_2} = \dfrac{1}{|z_2|} e^{-i\theta_2}$

3) $z_1 z_2 = |z_1||z_2| e^{i(\theta_1 + \theta_2)}$

4) $z_1 + z_2 = (|z_1| + |z_2|) e^{i(\theta_1 + \theta_2)}$

7.7. Calculate the value of $z\bar{z}$ if:

$$z = \frac{(1 + 2i)(1 + 3i)(1 + 4i)}{(2 - 3i)(2 - 4i)}$$

Difficulty level   ○ Easy   ● Normal   ○ Hard
Calculation amount   ● Small   ○ Normal   ○ Large

1) $\dfrac{86}{25}$

2) $\dfrac{96}{25}$

3) $\dfrac{85}{26}$

4) $\dfrac{95}{26}$

7.8. Calculate the value of i^i.

Difficulty level   ○ Easy   ● Normal   ○ Hard
Calculation amount   ● Small   ○ Normal   ○ Large

1) $e^{\frac{2}{\pi}}$

2) $e^{-\frac{\pi}{2}}$

3) $\text{Ln } i$

4) $\text{Ln } \pi$

7.9. Present the equation below in complex form.

$$x^2 - y^2 = 1$$

Difficulty level   ○ Easy   ● Normal   ○ Hard
Calculation amount   ○ Small   ● Normal   ○ Large

1) $(\bar{z})^2 - z^2 = 2$
2) $z^2 + (\bar{z})^2 = 2$
3) $z^2 - (\bar{z})^2 = 1$
4) $z^2 + (\bar{z})^2 = 1$

7.10. Calculate the value of following complex number if $n \in \mathbb{N}$.

$$\frac{(1+i)^n}{(1-i)^{n-2}}$$

Difficulty level   ○ Easy   ● Normal   ○ Hard
Calculation amount   ○ Small   ● Normal   ○ Large
1) $2i^{n-1}$
2) $\sqrt{2}i^2$
3) $\sqrt{2}i$
4) i^{-2}

7.11. Calculate the value of θ if the complex number below does not have a real part.

$$\frac{3 + 2i\sin\theta}{1 - 2i\sin\theta}$$

Difficulty level   ○ Easy   ● Normal   ○ Hard
Calculation amount   ○ Small   ● Normal   ○ Large
1) $\dfrac{\pi}{6}$
2) $\dfrac{\pi}{4}$
3) $\dfrac{\pi}{3}$
4) $\dfrac{\pi}{2}$

7.12. Calculate the value of following relation.

$$z = (1-i)(1 + i\sqrt{3})$$

Difficulty level   ○ Easy   ● Normal   ○ Hard
Calculation amount   ○ Small   ● Normal   ○ Large
1) $2\sqrt{2}\left(\cos\left(\dfrac{\pi}{12}\right) + i\sin\left(\dfrac{\pi}{12}\right)\right)$
2) $2\sqrt{2}\left(\cos\left(\dfrac{\pi}{12}\right) - i\sin\left(\dfrac{\pi}{12}\right)\right)$
3) $2\sqrt{2}\left(\cos\left(\dfrac{7\pi}{12}\right) + i\sin\left(\dfrac{7\pi}{12}\right)\right)$
4) $2\sqrt{2}\left(\cos\left(\dfrac{7\pi}{12}\right) - i\sin\left(\dfrac{7\pi}{12}\right)\right)$

7.13. Calculate the value of following relation.

$$\frac{(1+i)^{15}}{(1-i)^{13}}$$

Difficulty level   ○ Easy   ● Normal ○ Hard
Calculation amount   ○ Small ● Normal ○ Large
1) 2
2) −2
3) −3
4) 3

7.14. Calculate the value of relation below.

$$\left(\frac{\sqrt{3}+i}{\sqrt{3}-i}\right)^{10}$$

Difficulty level   ○ Easy   ● Normal ○ Hard
Calculation amount   ○ Small ● Normal ○ Large
1) $\dfrac{1+i\sqrt{3}}{2}$
2) $\dfrac{-1+i\sqrt{3}}{2}$
3) $\dfrac{1-i\sqrt{3}}{2}$
4) $\dfrac{-1-i\sqrt{3}}{2}$

7.15. Calculate the value of following relation.

$$\left(1+\cos\frac{2\pi}{3}+i\sin\frac{2\pi}{3}\right)^{120}$$

Difficulty level   ○ Easy   ● Normal ○ Hard
Calculation amount   ○ Small ● Normal ○ Large
1) −1
2) 1
3) −i
4) i

7.16. In the equation below, calculate the value of a.

$$e^{a+ib} = 1 - i\sqrt{3}$$

Difficulty level   ○ Easy   ● Normal ○ Hard
Calculation amount   ○ Small ● Normal ○ Large
1) ln 2
2) ln $\sqrt{2}$
3) 2
4) $\sqrt{2}$

7.17. Calculate the value of relation below if $z_m = \cos \dfrac{\pi}{2^m} + i \sin \dfrac{\pi}{2^m}$.

$$\prod_{m=1}^{\infty} z_m$$

Difficulty level   ○ Easy   ○ Normal   ● Hard
Calculation amount   ○ Small   ● Normal   ○ Large
1) $-\pi i$
2) πi
3) 1
4) -1

7.18. Which one of the following choices is correct if z_1 and z_2 are two non-zero complex numbers where:

$$\left| \frac{z_1 - \bar{z}_2}{z_1 + \bar{z}_2} \right| = 1$$

Difficulty level   ○ Easy   ○ Normal   ● Hard
Calculation amount   ○ Small   ● Normal   ○ Large
1) $Re(z_1 z_2) > 0$
2) $Re(z_1 z_2) < 0$
3) $Re(z_1 z_2) = 0$
4) $Im(z_1 z_2) > 0$

7.19. Calculate the maximum value of z if the relation below is held.

$$\left| \frac{6z - i}{2 + 3iz} \right| \leq 1$$

Difficulty level   ○ Easy   ○ Normal   ● Hard
Calculation amount   ○ Small   ● Normal   ○ Large
1) $\dfrac{1}{5}$

2) $\dfrac{1}{4}$

3) $\dfrac{1}{3}$

4) $\dfrac{1}{2}$

7.20. Calculate the value of the relation below in complex form.

$$\frac{1 + \sin \theta + i \cos \theta}{1 + \sin \theta + i \cos \theta}$$

Difficulty level   ○ Easy   ○ Normal   ● Hard
Calculation amount   ○ Small   ○ Normal   ● Large
1) $\sin \theta + i \cos \theta$
2) $\sin \theta - i \cos \theta$
3) $1 - i$
4) $\sin \dfrac{\theta}{2} - i \cos \dfrac{\theta}{2}$

References

1. Rahmani-Andebili, M. (2023). Calculus I (2nd Ed.) – Practice Problems, Methods, and Solutions, Springer Nature, 2021.
2. Rahmani-Andebili, M. (2021). Calculus – Practice Problems, Methods, and Solutions, Springer Nature, 2021.
3. Rahmani-Andebili, M. (2021). Precalculus – Practice Problems, Methods, and Solutions, Springer Nature, 2021.

Abstract

In this chapter, the problems of the seventh chapter are fully solved, in detail, step-by-step, and with different methods.

8.1. As we know from Euler' formula [1–3]:

$$e^{\theta i} = \cos\theta + i\sin\theta$$

Therefore:

$$e^{0i} = \cos 0 + i\sin 0 = 1 + 0i$$

$$e^{0i} = 1$$

Choice (2) is the answer.

In this problem, the rules below were used.

$$\cos 0 = 1$$

$$\sin 0 = 0$$

8.2. As we know from Euler' formula:

$$e^{\theta i} = \cos\theta + i\sin\theta$$

Therefore:

$$e^{\pi i} = \cos\pi + i\sin\pi = -1 + 0i$$

$$e^{\pi i} = -1$$

Choice (3) is the answer.

In this problem, the rules below were used.

$$\cos\pi = -1$$

$$\sin\pi = 0$$

8.3. The complex conjugate of a complex number can be achieved by changing the sign of imaginary part of the complex number while the other parts of the complex number are left intact. In other words:

$$\overline{(a+ib)} = (a+ib)^* = a - ib$$

Therefore:

$$\bar{i} = i^* = -i$$

Choice (4) is the answer.

8.4. All the relations are correct except Choice (3). Its correct relation is as follows.

$$(a+ib)(a-ib) = a^2 - aib + iba - (ib)^2$$

$$\Rightarrow (a+ib)(a-ib) = a^2 + b^2$$

Choice (3) is the answer.

In this problem, the rule below was used.

$$i^2 = -1$$

8.5. All the relations are correct except Choice (3). Its correct relation is as follows.

$$z_1 z_2 = (x_1 + iy_1)(x_2 + iy_2) = x_1 x_2 + x_1 iy_2 + iy_1 x_2 + i^2 y_1 y_2$$

$$\Rightarrow z_1 z_2 = (x_1 x_2 - y_1 y_2) + i(x_1 y_2 + y_1 x_2)$$

$$\Rightarrow Im(z_1 z_2) = x_1 y_2 + y_1 x_2$$

Choice (3) is the answer.

In this problem, the rule below was used.

$$i^2 = -1$$

8.6. All the relations are correct except Choice (4). Its correct relation is as follows.

$$z_1 + z_2 = |z_1|e^{i\theta_1} + |z_2|e^{i\theta_2}$$

Choice (4) is the answer.

8.7. Based on the information given in the problem, we have:

$$z = \frac{(1+2i)(1+3i)(1+4i)}{(2-3i)(2-4i)}$$

$$\Rightarrow \bar{z} = \frac{(1-2i)(1-3i)(1-4i)}{(2+3i)(2+4i)}$$

$$\Rightarrow z\bar{z} = \frac{(1+4)(1+9)(1+16)}{(4+9)(4+16)} = \frac{5 \times 10 \times 17}{13 \times 20}$$

$$\Rightarrow z\bar{z} = \frac{85}{26}$$

Choice (3) is the answer.

In this problem, the rules below were used.

$$\overline{(a+ib)} = a - ib$$

$$\overline{\left(\frac{z_1 z_2}{z_3 z_4}\right)} = \frac{\overline{z_1}\,\overline{z_2}}{\overline{z_3}\,\overline{z_4}}$$

$$(a+ib)(a-ib) = a^2 + b^2$$

8.8. The problem can be solved by transferring from Cartesian coordinate to polar coordinate as follows.

$$i = \sqrt{1}e^{i\tan^{-1}\frac{1}{0}} = e^{i\tan^{-1}\infty} = e^{\frac{\pi}{2}i}$$

Then:

$$i^i = \left(e^{\frac{\pi}{2}i}\right)^i = e^{\frac{\pi}{2}i^2} = e^{-\frac{\pi}{2}}$$

Choice (2) is the answer.

In this problem, the rules below were used.

$$a+ib = \sqrt{a^2+b^2}e^{i\tan^{-1}\left|\frac{b}{a}\right|} \quad \text{if } a>0, b>0$$

$$a+ib = \sqrt{a^2+b^2}e^{i\left(\pi - \tan^{-1}\left|\frac{b}{a}\right|\right)} \quad \text{if } a<0, b>0$$

$$a+ib = \sqrt{a^2+b^2}e^{i\left(\pi + \tan^{-1}\left|\frac{b}{a}\right|\right)} \quad \text{if } a<0, b<0$$

$$a+ib = \sqrt{a^2+b^2}e^{-i\tan^{-1}\left|\frac{b}{a}\right|} \quad \text{if } a>0, b<0$$

$$\left(e^{ia}\right)^b = e^{iab}$$

$$i^2 = -1$$

8.9. Based on the information given in the problem, we have:

$$x^2 - y^2 = 1$$

As we know:

$$z = x + iy \Rightarrow z^2 = (x+iy)^2 = x^2 - y^2 + i2xy$$

$$\bar{z} = x - iy \Rightarrow (\bar{z})^2 = x^2 - y^2 - i2xy$$

$$\Rightarrow z^2 + (\bar{z})^2 = 2\left(x^2 - y^2\right)$$

Therefore:

$$z^2 + (\bar{z})^2 = 2$$

Choice (2) is the answer.

In this problem, the rules below were used.

$$i^2 = -1$$

$$\overline{(a + ib)} = a - ib$$

8.10. Based on the information given in the problem, we have:

$$\frac{(1+i)^n}{(1-i)^{n-2}} \tag{1}$$

The problem can be solved as follows.

$$\Rightarrow \frac{(1+i)^n}{(1-i)^{n-2}} = \frac{(1+i)^{n-2}}{(1-i)^{n-2}}(1+i)^2 = \left(\frac{1+i}{1-i}\right)^{n-2}(1+2i-1) = \left(\frac{1+i}{1-i}\right)^{n-2}(2i) \tag{2}$$

On the other hand:

$$\left(\frac{1+i}{1-i}\right) = \frac{1+i}{1-i} \times \frac{1+i}{1+i} = \frac{1+2i+i^2}{1-i^2} = \frac{2i}{2} = i \tag{3}$$

Solving (1)–(3):

$$\frac{(1+i)^n}{(1-i)^{n-2}} = i^{n-2}(2i)$$

$$\Rightarrow \frac{(1+i)^n}{(1-i)^{n-2}} = 2i^{n-1}$$

Choice (1) is the answer.

In this problem, the rules below were used.

$$i^2 = -1$$

$$\frac{a+ib}{c+id} = \frac{a+ib}{c+id} \times \frac{c-id}{c-id}$$

8.11. Based on the information given in the problem, we know that:

$$Re\left(\frac{3 + 2i\sin\theta}{1 - 2i\sin\theta}\right) = 0 \tag{1}$$

The problem can be solved as follows.

$$\frac{3 + 2i\sin\theta}{1 - 2i\sin\theta} = \frac{3 + 2i\sin\theta}{1 - 2i\sin\theta} \times \frac{1 + 2i\sin\theta}{1 + 2i\sin\theta}$$

$$= \frac{3 + 6i\sin\theta + 2i\sin\theta - 4\sin^2\theta}{1 + 4\sin^2\theta} = \frac{(3 - 4\sin^2\theta) + 8i\sin\theta}{1 + 4\sin^2\theta}$$

$$= \frac{(3 - 4\sin^2\theta)}{1 + 4\sin^2\theta} + \frac{8\sin\theta}{1 + 4\sin^2\theta}i \tag{2}$$

Solving (1) and (2):

$$\frac{(3 - 4\sin^2\theta)}{1 + 4\sin^2\theta} = 0 \Rightarrow 3 - 4\sin^2\theta = 0 \Rightarrow \sin^2\theta = \frac{3}{4} \Rightarrow \sin\theta = \pm\frac{\sqrt{3}}{2} \Rightarrow \theta = \pm\frac{\pi}{3}$$

Choice (3) is the answer.

In this problem, the rules below were used.

$$\frac{a + ib}{c + id} = \frac{a + ib}{c + id} \times \frac{c - id}{c - id}$$

$$i^2 = -1$$

8.12. The problem can be solved by transferring from Cartesian coordinate to polar coordinate as follows.

$$1 - i = \sqrt{1 + (-1)^2}e^{-i\frac{\pi}{4}} = \sqrt{2}e^{-i\frac{\pi}{4}}$$

$$1 + i\sqrt{3} = \sqrt{(1)^2 + \left(\sqrt{3}\right)^2}e^{i\frac{\pi}{3}} = 2e^{i\frac{\pi}{3}}$$

Therefore:

$$z = (1 - i)\left(1 + i\sqrt{3}\right) = \left(\sqrt{2}e^{-i\frac{\pi}{4}}\right)\left(2e^{i\frac{\pi}{3}}\right) = 2\sqrt{2}e^{i\left(\frac{\pi}{3} - \frac{\pi}{4}\right)} = 2\sqrt{2}e^{i\frac{\pi}{12}}$$

$$\Rightarrow z = 2\sqrt{2}\left[\cos\left(\frac{\pi}{12}\right) + i\sin\left(\frac{\pi}{12}\right)\right]$$

Choice (1) is the answer.

In this problem, the rules below were used.

$$a + ib = \sqrt{a^2 + b^2}e^{i\tan^{-1}\left|\frac{b}{a}\right|} \quad \text{if } a > 0, b > 0$$

$$a + ib = \sqrt{a^2 + b^2}\, e^{i\left(\pi - \tan^{-1}\left|\frac{b}{a}\right|\right)} \quad \text{if } a < 0, b > 0$$

$$a + ib = \sqrt{a^2 + b^2}\, e^{i\left(\pi + \tan^{-1}\left|\frac{b}{a}\right|\right)} \quad \text{if } a < 0, b < 0$$

$$a + ib = \sqrt{a^2 + b^2}\, e^{-i\tan^{-1}\left|\frac{b}{a}\right|} \quad \text{if } a > 0, b < 0$$

$$\tan^{-1}(-1) = -\frac{\pi}{4}$$

$$\tan^{-1}\sqrt{3} = \frac{\pi}{3}$$

$$z_1 z_2 = \left(|z_1| e^{i\theta_1}\right)\left(|z_2| e^{i\theta_2}\right) = |z_1||z_2| e^{i(\theta_1 + \theta_2)}$$

$$|z| e^{i\theta} = |z|(\cos\theta + i\sin\theta)$$

8.13. The problem can be solved by transferring from Cartesian coordinate to polar coordinate as follows.

$$\frac{(1+i)^{15}}{(1-i)^{13}} = \frac{\left(\sqrt{2} e^{i\frac{\pi}{4}}\right)^{15}}{\left(\sqrt{2} e^{-i\frac{\pi}{4}}\right)^{13}}$$

$$= \frac{2 e^{i\frac{15\pi}{4}}}{e^{-i\frac{13\pi}{4}}} = 2 e^{i\left(\frac{15\pi}{4} - \left(-\frac{13\pi}{4}\right)\right)}$$

$$= 2 e^{i\frac{28\pi}{4}} = 2 e^{i7\pi}$$

$$= 2(\cos 7\pi + i\sin 7\pi) = 2(\cos\pi + i\sin\pi) = 2(-1 + 0i)$$

Thus:

$$\frac{(1+i)^{15}}{(1-i)^{13}} = -2$$

Choice (2) is the answer.

In this problem, the rules below were used.

$$a + ib = \sqrt{a^2 + b^2}\, e^{i\tan^{-1}\left|\frac{b}{a}\right|} \quad \text{if } a > 0, b > 0$$

$$a + ib = \sqrt{a^2 + b^2}\, e^{i\left(\pi - \tan^{-1}\left|\frac{b}{a}\right|\right)} \quad \text{if } a < 0, b > 0$$

$$a + ib = \sqrt{a^2 + b^2}\, e^{i\left(\pi + \tan^{-1}\left|\frac{b}{a}\right|\right)} \quad \text{if } a < 0, b < 0$$

$$a + ib = \sqrt{a^2 + b^2}\, e^{-i\tan^{-1}\left|\frac{b}{a}\right|} \quad \text{if } a > 0, b < 0$$

$$\tan^{-1}1 = \frac{\pi}{4}$$

$$\tan^{-1}(-1) = -\frac{\pi}{4}$$

$$\left(e^{ia}\right)^b = e^{iab}$$

$$\frac{z_1}{z_2} = \frac{|z_1|e^{i\theta_1}}{|z_2|e^{i\theta_2}} = \frac{|z_1|}{|z_2|}e^{i(\theta_1 - \theta_2)}$$

$$|z|e^{i\theta} = |z|(\cos\theta + i\sin\theta)$$

8.14. The problem can be solved by transferring from Cartesian coordinate to polar coordinate as follows.

$$\left(\frac{\sqrt{3}+i}{\sqrt{3}-i}\right)^{10} = \left(\frac{2e^{i\frac{\pi}{6}}}{2e^{-i\frac{\pi}{6}}}\right)^{10} = \left(e^{i\frac{\pi}{3}}\right)^{10} = e^{i\frac{10\pi}{3}}$$

$$= e^{i\left(2\pi+\frac{4\pi}{3}\right)} = e^{i\frac{4\pi}{3}}$$

$$= \cos\frac{4\pi}{3} + i\sin\frac{4\pi}{3} = -\frac{1}{2} - i\frac{\sqrt{3}}{2}$$

Therefore:

$$\left(\frac{\sqrt{3}+i}{\sqrt{3}-i}\right)^{10} = \frac{-1 - i\sqrt{3}}{2}$$

Choice (4) is the answer.

In this problem, the rules below were used.

$$a + ib = \sqrt{a^2 + b^2}\,e^{i\tan^{-1}\left|\frac{b}{a}\right|} \qquad \text{if } a>0, b>0$$

$$a + ib = \sqrt{a^2 + b^2}\,e^{i\left(\pi - \tan^{-1}\left|\frac{b}{a}\right|\right)} \qquad \text{if } a<0, b>0$$

$$a + ib = \sqrt{a^2 + b^2}\,e^{i\left(\pi + \tan^{-1}\left|\frac{b}{a}\right|\right)} \qquad \text{if } a<0, b<0$$

$$a + ib = \sqrt{a^2 + b^2}\,e^{-i\tan^{-1}\left|\frac{b}{a}\right|} \qquad \text{if } a>0, b<0$$

$$\tan^{-1}\frac{\sqrt{3}}{3} = \frac{\pi}{6}$$

$$\tan^{-1}\left(-\frac{\sqrt{3}}{3}\right) = -\frac{\pi}{6}$$

$$\frac{z_1}{z_2} = \frac{|z_1|e^{i\theta_1}}{|z_2|e^{i\theta_2}} = \frac{|z_1|}{|z_2|}e^{i(\theta_1 - \theta_2)}$$

$$\left(e^{ia}\right)^b = e^{iab}$$

$$e^{i(2\pi+\theta)} = e^{i\theta}$$

$$|z|e^{i\theta} = |z|(\cos\theta + i\sin\theta)$$

$$\cos \frac{4\pi}{3} = -\frac{1}{2}$$

$$\sin \frac{4\pi}{3} = -\frac{\sqrt{3}}{2}$$

8.15. The problem can be solved by transferring from Cartesian coordinate to polar coordinate as follows.

$$\left(1 + \cos \frac{2\pi}{3} + i \sin \frac{2\pi}{3}\right)^{120} = \left(1 - \frac{1}{2} + i \frac{\sqrt{3}}{2}\right)^{120}$$

$$= \left(\frac{1}{2} + i \frac{\sqrt{3}}{2}\right)^{120} = \left(e^{i\frac{\pi}{3}}\right)^{120} = e^{i40\pi} = e^{i(20\times 2\pi + 0)}$$

$$= \cos(0) + i \sin(0) = 1 + 0i$$

Therefore:

$$\left(1 + \cos \frac{2\pi}{3} + i \sin \frac{2\pi}{3}\right)^{120} = 1$$

Choice (2) is the answer.

In this problem, the rules below were used.

$$\cos \frac{2\pi}{3} = -\frac{1}{2}$$

$$\sin \frac{2\pi}{3} = \frac{\sqrt{3}}{2}$$

$$a + ib = \sqrt{a^2 + b^2}\, e^{i \tan^{-1}\left|\frac{b}{a}\right|} \qquad \text{if } a > 0, b > 0$$

$$a + ib = \sqrt{a^2 + b^2}\, e^{i\left(\pi - \tan^{-1}\left|\frac{b}{a}\right|\right)} \qquad \text{if } a < 0, b > 0$$

$$a + ib = \sqrt{a^2 + b^2}\, e^{i\left(\pi + \tan^{-1}\left|\frac{b}{a}\right|\right)} \qquad \text{if } a < 0, b < 0$$

$$a + ib = \sqrt{a^2 + b^2}\, e^{-i \tan^{-1}\left|\frac{b}{a}\right|} \qquad \text{if } a > 0, b < 0$$

$$\tan^{-1}\sqrt{3} = \frac{\pi}{3}$$

$$\left(e^{ia}\right)^b = e^{iab}$$

$$e^{i(2\pi + \theta)} = e^{i\theta}$$

$$|z| e^{i\theta} = |z|(\cos \theta + i \sin \theta)$$

$$\cos(0) = 1$$

$$\sin(0) = 0$$

8.16. Based on the information given in the problem, we have:

$$e^{a+ib} = 1 - i\sqrt{3} \tag{1}$$

The problem can be solved by transferring from Cartesian coordinate to polar coordinate as follows.

$$1 - i\sqrt{3} = 2e^{-\frac{\pi}{3}i} \tag{2}$$

As we know:

$$e^{a+ib} = e^a e^{ib} \tag{3}$$

Solving (1)–(3):

$$e^a e^{ib} = 2e^{-\frac{\pi}{3}i} \Rightarrow \begin{cases} e^a = 2 \\ e^{ib} = e^{-\frac{\pi}{3}i} \end{cases} \Rightarrow a = \ln 2, \quad b = -\frac{\pi}{3}$$

Choice (1) is the answer.

In this problem, the rules below were used.

$$a + ib = \sqrt{a^2 + b^2}\, e^{i\tan^{-1}\left|\frac{b}{a}\right|} \quad \text{if } a > 0, b > 0$$

$$a + ib = \sqrt{a^2 + b^2}\, e^{i\left(\pi - \tan^{-1}\left|\frac{b}{a}\right|\right)} \quad \text{if } a < 0, b > 0$$

$$a + ib = \sqrt{a^2 + b^2}\, e^{i\left(\pi + \tan^{-1}\left|\frac{b}{a}\right|\right)} \quad \text{if } a < 0, b < 0$$

$$a + ib = \sqrt{a^2 + b^2}\, e^{-i\tan^{-1}\left|\frac{b}{a}\right|} \quad \text{if } a > 0, b < 0$$

$$\tan^{-1}\left(-\sqrt{3}\right) = -\frac{\pi}{3}$$

$$e^{a+b} = e^a e^b$$

8.17. Based on the information given in the problem, we have:

$$z_m = \cos\frac{\pi}{2^m} + i\sin\frac{\pi}{2^m}$$

Therefore:

$$\prod_{m=1}^{\infty} z_m = \prod_{m=1}^{\infty} e^{\frac{\pi}{2^m}i} = e^{\frac{\pi}{2}i} e^{\frac{1}{2}\left(\frac{\pi}{2}\right)i} e^{\frac{1}{4}\left(\frac{\pi}{2}\right)i} \cdots = e^{\frac{\pi}{2}i + \frac{1}{2}\left(\frac{\pi}{2}i\right) + \frac{1}{4}\left(\frac{\pi}{2}i\right) + \cdots}$$

$$\Rightarrow \prod_{m=1}^{\infty} z_m = \exp\left(\sum_{m=1}^{\infty} \frac{\pi}{2^m}i\right) = e^{\sum_{m=1}^{\infty} \frac{\pi}{2^m}i} = e^{\frac{\frac{\pi}{2}i}{1-\frac{1}{2}}}$$

$$\Rightarrow \prod_{m=1}^{\infty} z_m = e^{\pi i} = \cos\pi + i\sin\pi = -1 + 0i$$

$$\Rightarrow \prod_{m=1}^{\infty} z_m = -1$$

Choice (4) is the answer.

In this problem, the rules below were used.

$$\prod_{m=1}^{\infty} z_m = z_1 z_2 \ldots z_\infty$$

$$e^{ix} = \cos x + i \sin x$$

$$\left(|z_1|e^{i\theta_1}\right)\left(|z_2|e^{i\theta_2}\right) = |z_1||z_2|e^{i(\theta_1+\theta_2)}$$

$$\sum_{m=1}^{\infty} a_m = \frac{a_1}{1-q} \qquad \text{if } |q| < 1$$

$$e^{\pi i} = -1$$

8.18. Based on the information given in the problem, we have:

$$\left|\frac{z_1 - \bar{z}_2}{z_1 + \bar{z}_2}\right| = 1 \tag{1}$$

Let us assume:

$$z_1 = x_1 + iy_1 \tag{2}$$

$$z_2 = x_2 + iy_2 \tag{3}$$

Solving (1)–(3):

$$\left|\frac{(x_1 + iy_1) - (x_2 - iy_2)}{(x_1 + iy_1) + (x_2 - iy_2)}\right| = 1 \Rightarrow \left|\frac{(x_1 - x_2) + i(y_1 + y_2)}{(x_1 + x_2) + i(y_1 - y_2)}\right| = 1$$

$$\Rightarrow \frac{|(x_1 - x_2) + i(y_1 + y_2)|}{|(x_1 + x_2) + i(y_1 - y_2)|} = 1$$

$$\Rightarrow |(x_1 - x_2) + i(y_1 + y_2)| = |(x_1 + x_2) + i(y_1 - y_2)|$$

$$\Rightarrow \sqrt{(x_1 - x_2)^2 + (y_1 + y_2)^2} = \sqrt{(x_1 + x_2)^2 + (y_1 - y_2)^2}$$

$$\Rightarrow (x_1)^2 + (x_2)^2 - 2x_1x_2 + (y_1)^2 + (y_2)^2 + 2y_1y_2 = (x_1)^2 + (x_2)^2 + 2x_1x_2 + (y_1)^2 + (y_2)^2 - 2y_1y_2$$

$$\Rightarrow -4x_1x_2 + 4y_1y_2 = 0 \Rightarrow x_1x_2 - y_1y_2 = 0$$

$$\Rightarrow Re(z_1z_2) = 0$$

Choice (3) is the answer.

In this problem, the rules below were used.

$$\left|\frac{a+ib}{c+id}\right| = \frac{|a+ib|}{|c+id|}$$

$$|a+ib| = \sqrt{a^2+b^2}$$

$$Re(z_1 z_2) = Re((x_1+iy_1)(x_2+iy_2)) = x_1 x_2 - y_1 y_2$$

$$i^2 = -1$$

8.19. Based on the information given in the problem, we have:

$$\left|\frac{6z-i}{2+3iz}\right| \leq 1$$

$$\frac{|6z-i|}{|2+3iz|} \leq 1 \Rightarrow |6(x+iy)-i| \leq |2+3i(x+iy)|$$

$$\Rightarrow |6x+i(6y-1)| \leq |(2-3y)+3ix|$$

$$\Rightarrow \sqrt{(6x)^2+(6y-1)^2} \leq \sqrt{(2-3y)^2+(3x)^2}$$

$$\Rightarrow 36x^2+36y^2-12y+1 \leq 4-12y+9y^2+9x^2$$

$$27x^2+27y^2 \leq 3 \Rightarrow x^2+y^2 \leq \frac{1}{9} \Rightarrow \sqrt{x^2+y^2} \leq \frac{1}{3}$$

$$\Rightarrow |z| \leq \frac{1}{3}$$

$$\Rightarrow \max(|z|) = \frac{1}{3}$$

Choice (3) is the answer.

In this problem, the rules below were used.

$$\left|\frac{a+ib}{c+id}\right| = \frac{|a+ib|}{|c+id|}$$

$$|a+ib| = \sqrt{a^2+b^2}$$

$$i^2 = -1$$

8.20. Based on the information given in the problem, we have:

$$\frac{1+\sin\theta+i\cos\theta}{1+\sin\theta-i\cos\theta}$$

The problem can be solved as follows.

$$\frac{1 + \sin\theta + i\cos\theta}{1 + \sin\theta - i\cos\theta} = \frac{1 + \sin\theta + i\cos\theta}{1 + \sin\theta - i\cos\theta} \times \frac{1 + \sin\theta + i\cos\theta}{(1 + \sin\theta) + i\cos\theta}$$

$$= \frac{(1 + \sin\theta + i\cos\theta)^2}{(1 + \sin\theta)^2 - i^2\cos^2\theta} = \frac{1 + \sin^2\theta + (i\cos\theta)^2 + 2\sin\theta + 2\sin\theta i\cos\theta + 2i\cos\theta}{1 + \sin^2\theta + 2\sin\theta + \cos^2\theta}$$

$$= \frac{1 + \sin^2\theta - \cos^2\theta + 2\sin\theta + 2i\cos\theta(1 + \sin\theta)}{2 + 2\sin\theta}$$

$$= \frac{\sin^2\theta + \cos^2\theta + \sin^2\theta - \cos^2\theta + 2\sin\theta + 2i\cos\theta(1 + \sin\theta)}{2 + 2\sin\theta}$$

$$= \frac{2\sin^2\theta + 2\sin\theta + 2i\cos\theta(1 + \sin\theta)}{2 + 2\sin\theta}$$

$$= \frac{2\sin\theta(1 + \sin\theta) + 2i\cos\theta(1 + \sin\theta)}{2(1 + \sin\theta)}$$

$$= \frac{2(1 + \sin\theta)(\sin\theta + i\cos\theta)}{2(1 + \sin\theta)} = \sin\theta + i\cos\theta$$

Therefore:

$$\frac{1 + \sin\theta + i\cos\theta}{1 + \sin\theta - i\cos\theta} = \sin\theta + i\cos\theta$$

Choice (1) is the answer.

In this problem, the rules below were used.

$$\frac{a + ib}{c + id} = \frac{a + ib}{c + id} \times \frac{c - id}{c - id}$$

$$(a + ib)(a - ib) = a^2 + b^2$$

$$(a + b + c)^2 = a^2 + b^2 + c^2 + 2ab + 2bc + 2ac$$

$$\sin^2\theta + \cos^2\theta = 1$$

References

1. Rahmani-Andebili, M. (2023). Calculus I (2nd Ed.) – Practice Problems, Methods, and Solutions, Springer Nature, 2021.
2. Rahmani-Andebili, M. (2021). Calculus – Practice Problems, Methods, and Solutions, Springer Nature, 2021.
3. Rahmani-Andebili, M. (2021). Precalculus – Practice Problems, Methods, and Solutions, Springer Nature, 2021.

Index

M. Rahmani-Andebili, *Calculus II*, https://doi.org/10.1007/978-3-031-45353-3